Color TV Case Histories Illustrated: Photo Guide to Troubles & Cures—Vol. 2

Color TV Case Histories Illustrated: Photo Guide to Troubles & Cures—Vol. 2

By Robert L. Goodman

TAB BOOKS
Blue Ridge Summit, Pa. 17214

FIRST EDITION

FIRST PRINTING— JULY 1977

Printed in the United States
of America

Library of Congress Cataloging in Publication Data (Revised)

Goodman, Robert L
Color TV case histories illustrated.

Subtitle varies on v. 2: Photo guide to trouble-shooting.
Includes indexes.
1. Color television--Repairing. I. Title.
TK6670.G55 621.3888'7 74-33619
ISBN 0-8306-5746-0 (v. 1)
ISBN 0-8306-4746-5 pbk. (v. 1)

Cover photo courtesy "Electronic Servicing"

Preface

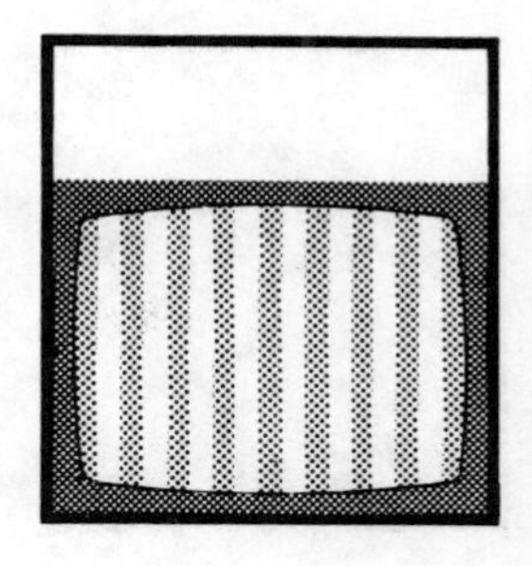

The TV trouble symptoms illustrated in this book are all true case-history problems. The professional TV technician may see a few "old friends" or enemies, as the case may be, among the symptoms included, even though an attempt has been made to eliminate very obvious problems.

It has been said that a picture is worth a thousand words. I believe you will find the picture symptom photos in this second volume of case histories will help you locate circuit faults more quickly and efficiently than reading many pages of words. Of course, words concerning circuit explanation and troubleshooting techniques are essential. So the photos are accompanied by a circuit schematic and a simplified discussion of how and why the system functions. After these initial particulars, I have included some helpful hints (probable faults) that you may use in locating similar troubles.

I want to thank the following color TV manufacturers and their representatives for their schematics, circuit information, and assistance in preparing this volume:

Admiral (Rockwell Corporation) Richard Keeran
General Electric—Frank Boston and Bob Movak
The Magnavox Company—Ray Leranko
Quasar Inc. (Motorola)—Edward Amaitis, El Mueller
Philco-Ford—John R. Krawczyk, Fred Fischer
RCA Sales Corporation—Herb Horton, G.F. Corne
Sylvania—E. M. Nanni
Zenith Radio Corporation—Brian Marohnic, Ed Kob, Ed Krol, Leo Smith, Joe Barrett, and Charles (Pinky) Osborne

Bob Goodman

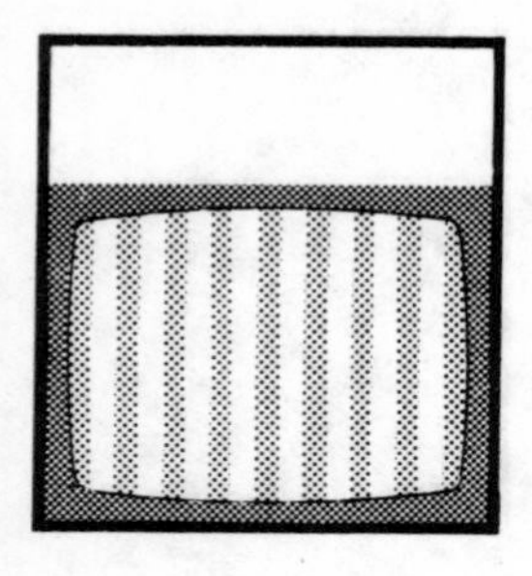

About This Book

Color symptoms included here are arranged by their obvious visual classification. The second section, *Weak or Snowy Picture*, for example, encompasses those problems that appear to be traceable to insufficient RF reaching the receiver. The symptom photos show actual sets caught in the act of being incorrigible in this special way. But as you read the text and start tracking down problems in sets having comparable problems, you'll soon learn that not all weak-picture symptoms point to antenna or tuner problems.

Within each section, individual case-history picture symptoms are shown. In most cases, the customer's TV set is identified by make and model. This is followed by a verbal description of the symptom, often with variations that are likely to occur. The analysis then tells where to start looking for the problem and gives the technical details of the circuit in question. The final paragraph gives the probable causes of such symptoms, including the specific cause of the documented problem.

We hope this book will prove irreplaceably serviceable to you in your work, and that it will save you many hours of tedious—and costly—troubleshooting.

THE EDITORS

Other TAB Books by Robert L. Goodman

Book No. 436	Practical Color TV Servicing Techniques—2nd Edition
Book No. 502	Zenith Color TV Service Manual—Vol. 1
Book No. 522	Philco Color TV Manual
Book No. 536	General Electric Volume 1
Book No. 562	Zenith Color TV Service Manual—Vol. 2
Book No. 577	How to Use Color TV Test Instruments
Book No. 595	199 Color TV Troubles & Solutions
Book No. 633	Simplified TV Trouble Diagnoisis
Book No. 668	Zenith Color TV Service Manual—Vol. 3
Book No. 696	TV Tuner Schematic/Servicing Manual
Book No. 706	Indexed Guide to Modern Electronic Circuits
Book No. 746	Color TV Case Histories Illustrated: Photo Guide to Troubles & Cures—Vol. 1
Book No. 772	Troubleshooting with the Dual-Trace Scope
Book No. 838	Zenith Color TV Service Manual—Vol. 4

Contents

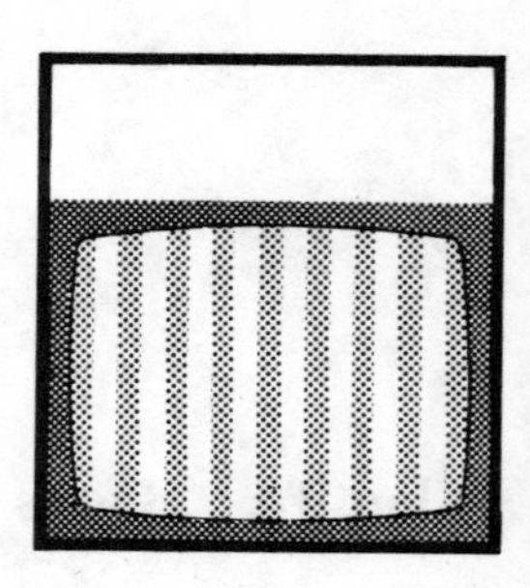

Color TV Case Histories Illustrated: Photo Guide to Troubles & Cures–Vol. 2

Introduction to Simplified TV Picture Diagnosis

Question: How do you diagnose color TV troubles if you do not have years of troubleshooting experience under your belt and have not graduated from the electronics college of hard knocks? Is there an article or electronics book that you could read and study that would make you an ace color-TV troubleshooter in a few weeks?

Answer: With my years of observations I would say there is no sure-fire way—you just cannot beat years of actual experience along with constant study to keep abreast with new additions to the state of the art. But what we will try to do in this book is to get the newer, inexperienced technician off to a good start on ways to quickly diagnose color TV symptoms and problems. And this may also be of value as a review to the old timers.

When you have a color set to repair and the screen produces strange symptoms, it is a time to make some decisions. You are at the crossroads—which way do you go now? First, it would be wise to stop, look, and listen—sometimes to smell or touch. Then think. Use the thinking technician's approach to circuit diagnosis.

No doubt about it, a picture is worth a thousand words, or more, and photos of the TV screen symptoms will be fully

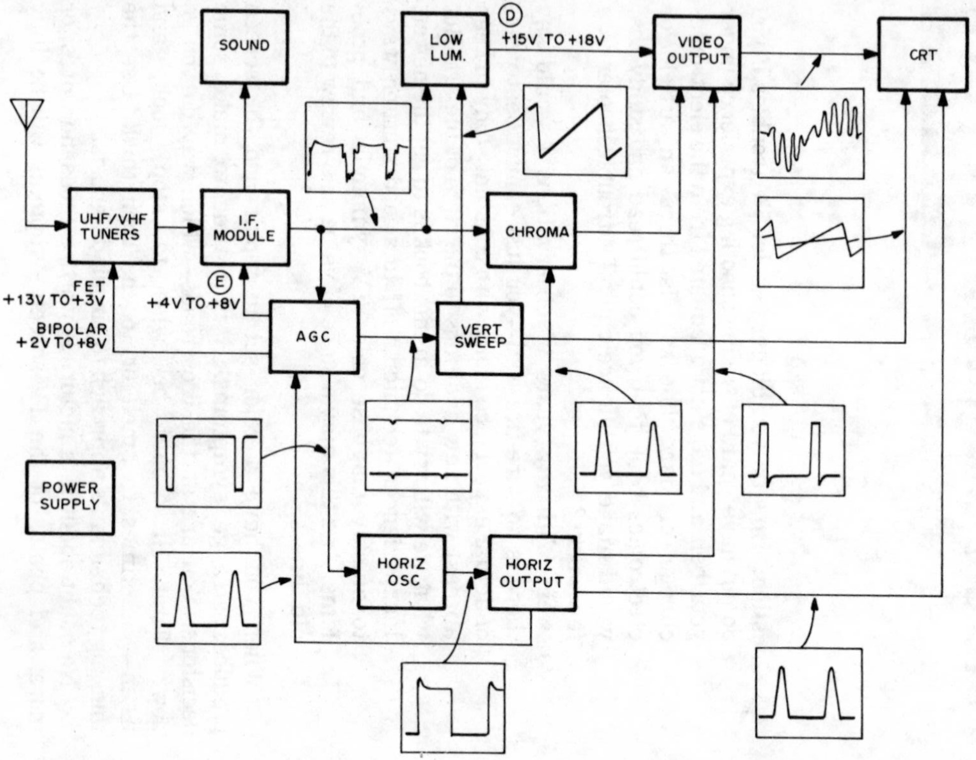

SOUND

LOW LUM.

D

+15V TO +18V

VIDEO OUTPUT

CRT

UHF/VHF TUNERS

I.F. MODULE

CHROMA

E

FET +13V TO +3V

BIPOLAR +2V TO +8V

+4V TO +8V

AGC

VERT SWEEP

POWER SUPPLY

HORIZ OSC

HORIZ OUTPUT

Fig. 1-1. Color TV block diagram.

utilized in this book. One of the best test instruments to use (and it's free) is to look at the picture, or lack of any picture, on the set's screen in question. Then try to visualize which section of the color chassis is at fault.

Now, follow along with the block diagram in Fig. 1-1 and let's look at a few color TV trouble examples.

BASIC PICTURE SYMPTOMS AND CAUSES

An effective way to troubleshoot any complex electronic circuitry is to break it down into small individual sections or circuits. Now let's put to use the where-to-look-for-trouble method when different picture symptoms appear.

No Picture or Sound (Dead on Arrival)

Check for blown fuses and tripped-out circuit breakers. Is the on/off switch good? How about the AC plug and power-cord interlock device? Next, check for AC voltages to and from the power transformers, and then for proper DC B+ voltages. Look for short circuits or an overload condition in B+ voltages, especially if fuses are blown. Prime suspects are shorted diode rectifiers in the power supply. Some sets use bridge diode circuits. Open filter capacitors in the power supply will cause a picture bend (Fig. 1-2) or a small picture raster that will pull in from top, bottom, and sides.

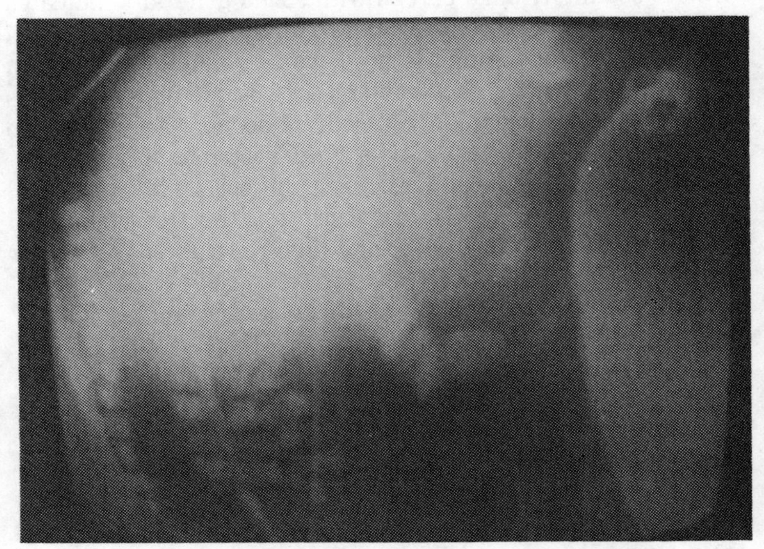

Fig. 1-2. Faulty filter capacitors cause bend in picture.

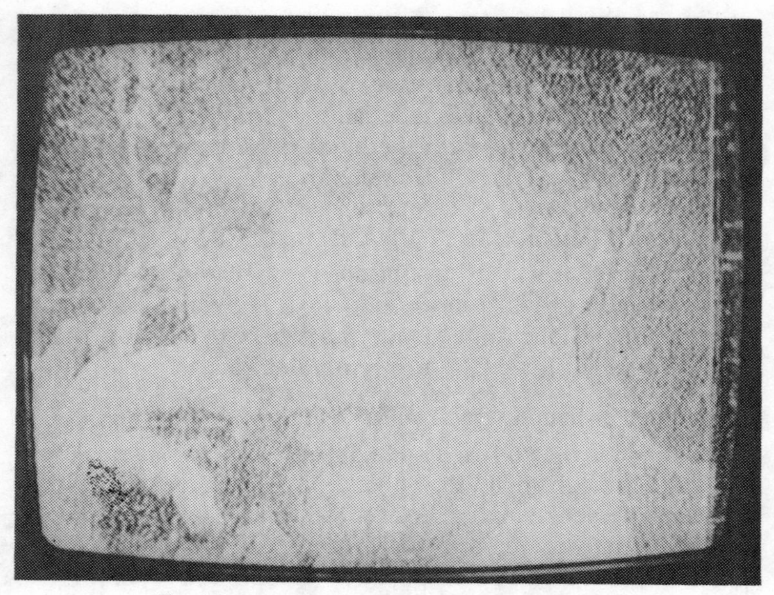

Fig. 1-3. Snowy picture—tuner or AGC problem.

Dark Screen With or Without Sound

Is the screen dark or does it have a bright raster with or without sound? A dark screen with sound generally indicates loss of horizontal sweep or high voltage to the CRT, or a defective CRT or wrong picture tube bias voltages. A screen that lights up with a good raster tells you the high voltage is okay, but the set has video trouble if the sound is good, while no sound points to video IF, AGC, or tuner system malfunctions.

For loss of high voltage, check the horizontal oscillator, horizontal output tube (transistor), horizontal sweep transformer, deflection yoke, and high-voltage (HV) rectifier tube. Late model color chassis use solid-state "stick" rectifiers or triplers. Also, do not overlook the damper and HV regulator tubes.

Snowy Picture

A weak, snowy picture (Fig. 1-3) usually points to problems in the tuner or to the AGC voltage that controls the gain of the tuner's RF amplifier stage. The tuner may only need new tubes and a good cleaning to correct this problem. If the picture flickers and flashes when you wiggle the channel knob, the tuner contacts are dirty and need to be cleaned. And do not overlook antenna or cable trouble.

A blank white screen with no sound lets you know that the oscillator or mixer stage in the tuner is at fault. Use a tuner from another set or a universal substitution tuner, if you have one, to verify this diagnosis.

Blank White Screen

A blank white screen with no sound (Fig. 1-4) could be caused by a faulty video-IF amplifier stage or an AGC voltage control problem. Use a bias-voltage substitution box to apply correct AGC voltage to the IF amplifiers. A very dark picture with too much contrast (Fig. 1-5), which may also bend or twist, is usually an AGC overload problem. If correct AGC bias voltage restores proper picture operation, the AGC system must then be checked out.

A blank white screen (or dark screen with good sound) and correct high voltage at the CRT anode would mean a fault in the video amplifiers or CRT bias circuits. Also, and this is easy to overlook, in some solid-state sets a faulty chroma demodulator IC will cause the same symptom.

Fig. 1-4. Blank white screen—video troubles.

Fig. 1-5. Very dark picture—AGC overload problem.

Picture Streaks

Dark lines or light streaks across the picture (Fig. 1-6) is usually caused by arcs and sparks within the color chassis. Look for arcs due to:

- Dust or moisture under HV anode cup on CRT
- Poor ground-strap connections around CRT
- Focus divider resistor
- Arcs within damper tube
- Break down of spark-gap protection devices
- Faulty focus or boost diode
- Focus connection (pin 9) or CRT
- Solid-state HV tripler or stick type rectifiers
- B+ bypass capacitors
- Arcs within any transformer
- Poor solder connections
- Loose plug-in modules and other cable plugs

LOSS OF COLOR, WEAK COLOR, OR WRONG COLORS

These faults are usually found in the chroma system. The chroma system is made up of the color killer, chroma amplifier circuit, AFC control stage, 3.58 MHz oscillator, and color demodulator circuits.

Preliminary Checks

Make sure the fine tuning is correct. Adjust the color-killer level control, and tune 3.58 MHz color oscillator for a zero beat. One common color trouble may show up as a "barber pole" affect—note color-bar pattern in Fig. 1-7. Refer to the color AFC and 3.58 MHz oscillator circuits in Fig. 1-8.

Color Circuit Trouble Comments and Checks

A properly operating AFC circuit will measure about zero volts with a VTVM at test point W under a no-signal condition.

Some possible voltages in excess of one volt at test point W under no-signal conditions can be caused by the following faults:

- A defective AGC phase detector diode (X12, X13). One diode may be conducting much more than the other, or be shorted or open.
- A faulty 3.58 MHz CW oscillator and control tube (V15). Oscillator operating off frequency.
- An open L45 burst transformer.

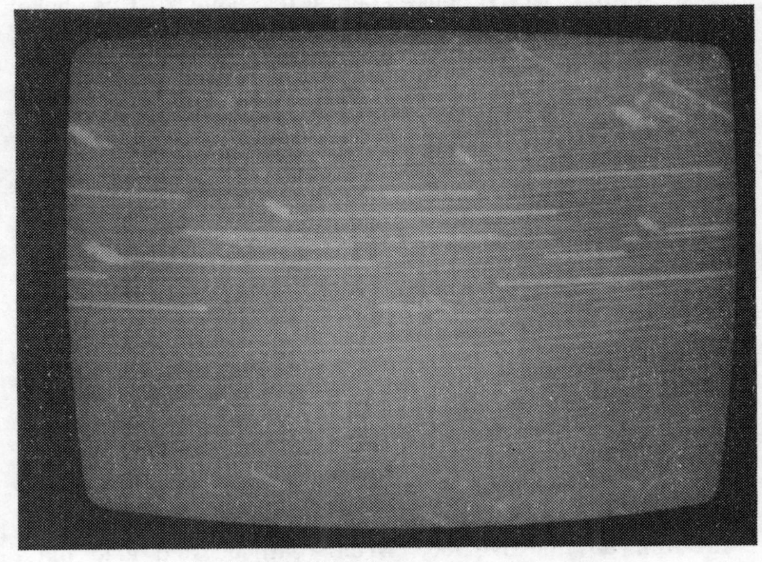

Fig. 1-6. Streaks across screen—internal arc.

Fig. 1-7. Candy-cane color stripe—loss of color sync.

- An open or leaky section in one of the dual 0.001 μF capacitors in the AFC phase detector circuit (C165, C166).
- A faulty 80 μH choke (146) at test point W. This choke is needed for oscillator stability.
- Component leakage or value change in the anti-hunt network. This circuit is located between test point W and ground.
- The 2.2 meg resistors may have changed value (not matched) in the AFC phase detector circuit. These must be replaced with a matched pair.
- Incorrect setting of the 3.58 MHz color-frequency oscillator coil (L47). Ground test point W and adjust 3.58 MHz oscillator for zero beat.

Loss of color could be (but is usually not) due to faults or misalignment in the tuner, IF amplifier system, or chroma amplifier stages. Loss of one color or wrong color trouble can generally be located in the color demodulator stage. Use a scope to check for correct phase of 3.58 MHz CW drive to the demodulators.

Again, do not overlook wrong cathode and grid bias voltages for the CRT, or a defective picture tube.

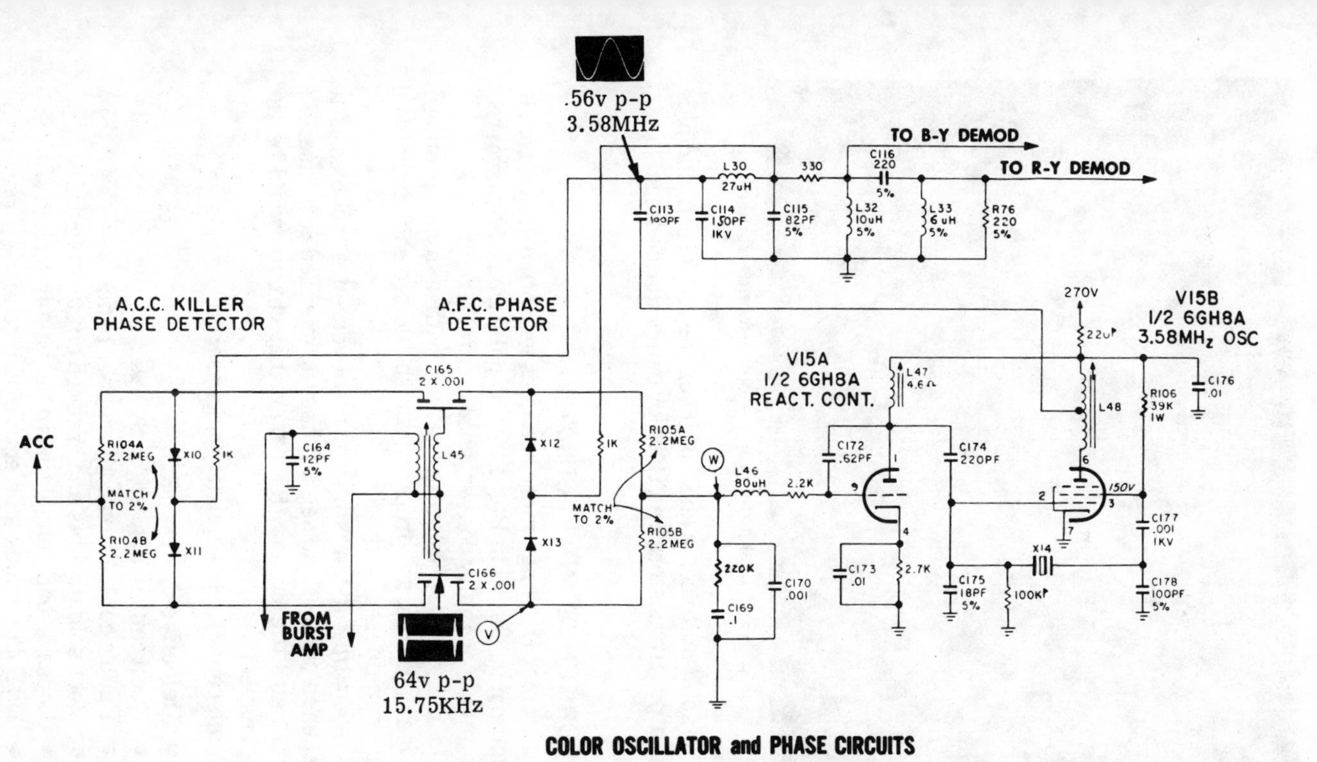

Fig. 1-8. Color sync and oscillator circuits.

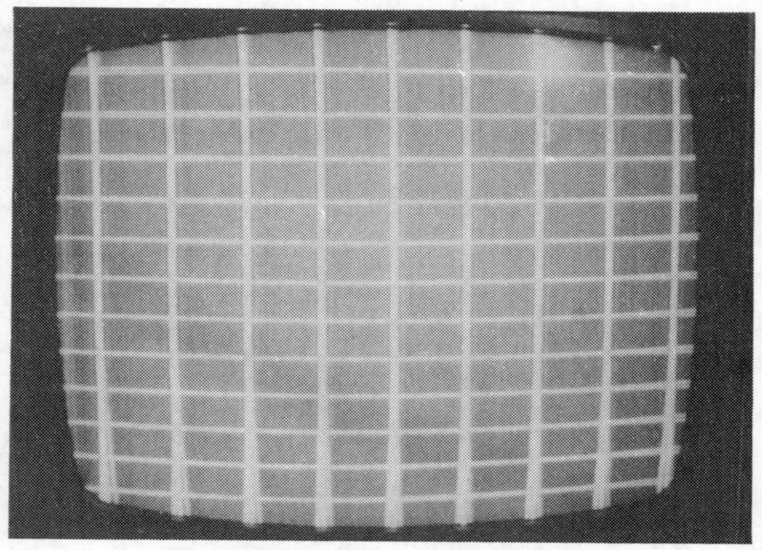

Fig. 1-9. Crosshatch pattern—convergence trouble.

For solid-state color chassis check the red, green, and blue video output transistors and IC color demodulator chip.

CONVERGENCE AND FOCUS TIPS

Any color fringing (more readily seen with a crosshatch or dot pattern as shown in Fig. 1-9) is usually caused by incorrect adjustments or faulty convergence control circuits.

Do not confuse misconvergence with a focus problem. With focus control set properly you should see each individual raster scan line.

A good tip for tracking down convergence troubles is to adjust each control and see if it has the intended response. If a control has little or no effect, then that circuit would be the prime suspect.

Some convergence quick checks are:

- Defective diodes.
- Open coils in convergence yoke.
- Faulty controls on convergence panel.
- Poor solder joints and cracked circuit boards.
- Loose or bad connection in the plug-in sockets.
- Do not overlook a defective picture tube (very rare).

If the raster scan line will not sharpen up, some probable component faults are:

- Bad focus rectifier tube.
- Faulty focus stick diode.
- Defective focus divider resistor.
- Some types of HV triplers that supply the focus voltage.
- Defective focus adjustment control or transformer.
- High voltage at CRT anode may be wrong.
- The focus pin (number nine) on the CRT socket may be defective.
- Spark gaps in the focus circuit may be shorted.

INTERMITTENT TROUBLES

Intermittent circuit faults are hard to locate in many chassis. Lots of times you will just have to play it by ear. The following is a check list to use for intermittents:

- Try to divide the set's circuits down to a section that is most likely to have the intermittent fault. Do this by gently tapping or probing around various parts of the chassis.
- Use test instruments (scopes, meters, and signal tracers) to monitor various circuit signals.
- Tubes and transistors may act up after they have generated enough internal heat.
- Capacitors become leaky after a long warmup.
- Coils may open when cold and "heal" themselves after warmup.
- Resistor values will usually change value over a long period of time.
- Use a Variac to increase the AC line voltage under test to help speedup the intermittent fault. This may help make the faulty component break down.
- Use a coolant spray, heat lamp, or gun to induce component breakdown. This will also help you locate the area of the fault.
- Carefully inspect all chassis connections for poor (cold) solder connections.

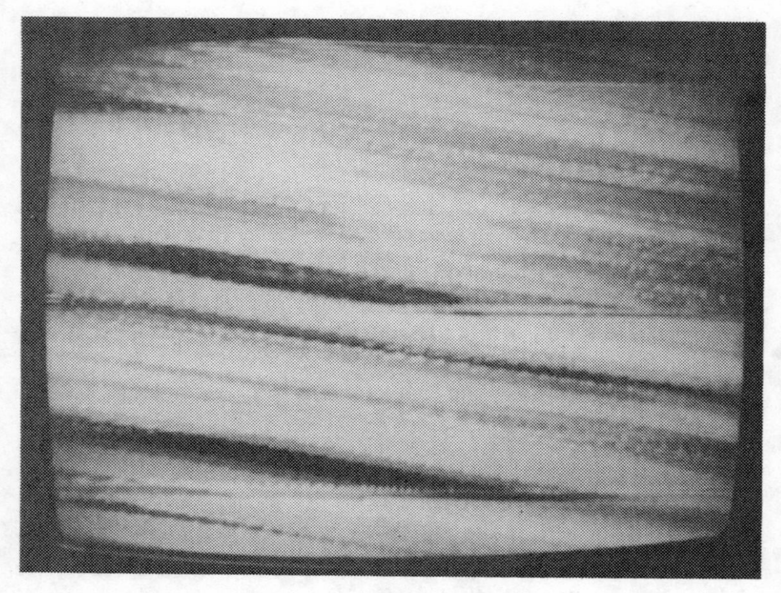

Fig. 1-10. Horizontal oscillator off frequency—loss of horizontal sync.

HORIZONTAL TROUBLES

If you came across the picture symptom shown in Fig. 1-10, where would you look for the problem? Let's analyze the picture symptoms and see which section is at fault.

When you see several slanting dark lines as in Fig. 1-10, this indicates a horizontal oscillator sweep, horizontal sync, or AFC (automatic frequency control) circuit problem. The dark lines can slope in either direction, depending upon whether the horizontal oscillator is too high or too low in frequency. Other times the picture may tend to roll from one side of the screen to the other, and appear to float by. This symptom is usually a horizontal sync problem.

Circuit Operation

Refer to the horizontal AFC and oscillator circuit in Fig. 1-11. This circuit has a pair of diodes used as an AFC phase detector, which accepts a pulse from the sweep output circuit and compares its phase (timing) with the incoming horizontal sync pulse. As the DC voltage output from the AFC phase detector goes more negative, the bias on the control tube increases, resulting in a higher frequency output from the oscillator. As the DC voltage from the phase detector goes

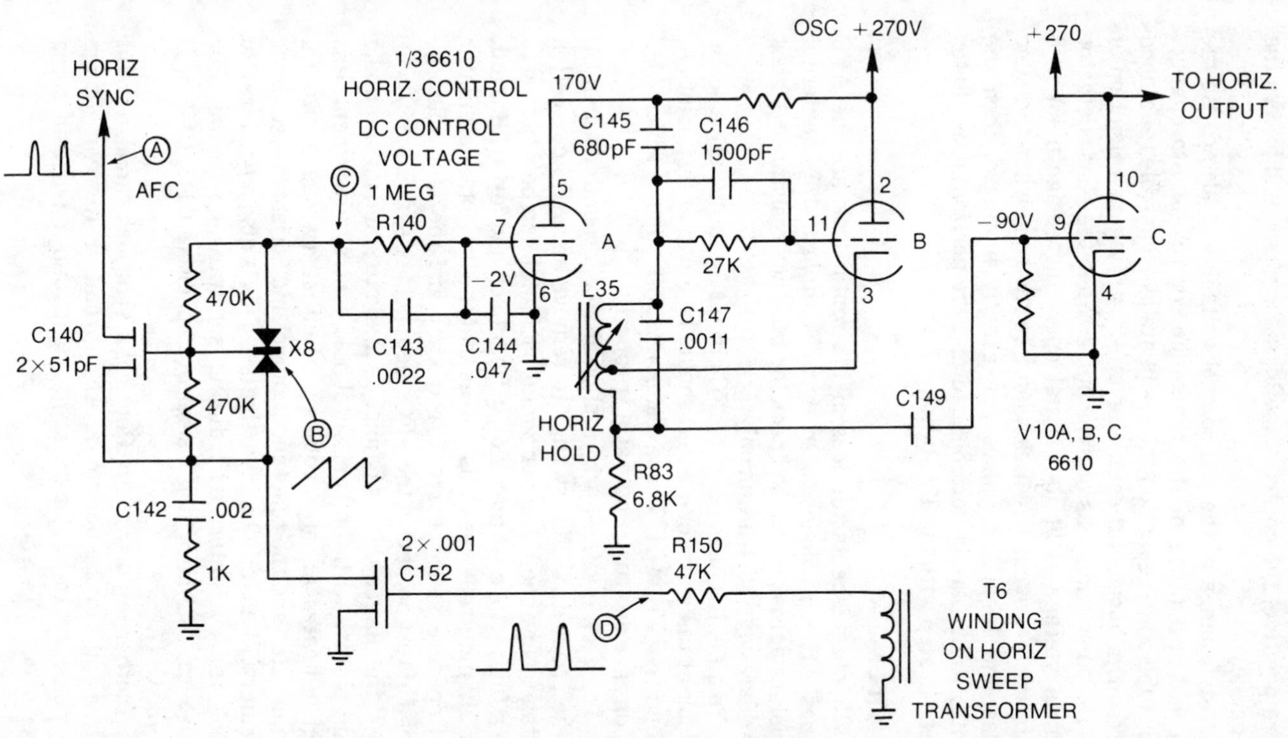

Fig. 1-11. Horizontal AFC and oscillator circuit.

more positive, the control tube bias decreases and lowers the frequency.

Any change of the DC control voltage will also cause the internal resistance of the reactance control tube to change. This tube control section then appears as a variable reactance across the horizontal oscillator tank, varying as a function of the DC correction bias or sync pulse. When the phase detector control voltage is at its normal operating design point (−2 volts in this circuit), which would be the case when the sync pulse and feedback pulses are exactly in phase, then the control tube bias is not changed and the horizontal oscillator keeps in step with the TV station.

Circuit Checks

Prime suspects for loss of horizontal lock are the AFC diodes. These diodes also have *matched* 470K external resistors across them, so check for matched resistance. Check the diodes by substitution only.

To quickly check out the AFC circuit for faults, use a scope to look at the waveforms that should be found at test points A, B, and D of Fig. 1-11. These test points will indicate which AFC actions are or are not operating properly.

To check out the horizontal oscillator, ground test point C to the chassis (with test clip) to remove any AFC correction voltage. Now see if the horizontal oscillator can be adjusted to see a full picture (it may slowly float back and forth). If a full, upright picture is seen this proves that the horizontal oscillator will operate at the correct frequency and the fault must be ahead of test point C in the AFC, feedback, or sync stages.

If the diagonal lines cannot be made to stand up and make a full picture with the horizontal hold control adjustment, this then indicates an off-frequency condition (wrong horizontal sweep rate) and the reactance control and horizontal oscillator circuits will have to be investigated. This is also a good way to check for oscillator drift during set warm-up period. For horizontal oscillator drift, suspect capacitors C145, C146, and maybe C147.

Another important tip. Poor color sync lock can also result from a fault in the horizontal AFC circuit, as this can change the phase of the color-burst gating (timing) pulse in some types of color TV sets.

VERTICAL SWEEP TIPS

When the sets picture rolls up or down (fast or slow) and cannot be stopped with the vertical hold control, look for trouble in the vertical oscillator circuit. A vertical roll symptom means that the vertical oscillator is not operating at the correct frequency.

If the screen cannot be filled out at the top or bottom (even with height and linearity controls at maximum setting), the defect is probably in the vertical output stage or deflection yoke.

A thin horizontal white line (Fig. 1-12) across the screen indicates no vertical sweep deflection at all. This could be caused by a dead vertical oscillator, faulty vertical output stage, vertical output transformer, or deflection yoke.

A picture that will not lock (slowly rolls up or down) is due to a loss of vertical sync pulses to the vertical oscillator stage. Check out the vertical sync circuits. For loss of both vertical and horizontal locking action, go to the sync clipper and amplifier stages.

PLUG-IN MODULES

If a suspected plug-in module is changed out and there is no change in the set's operating symptoms, the module is

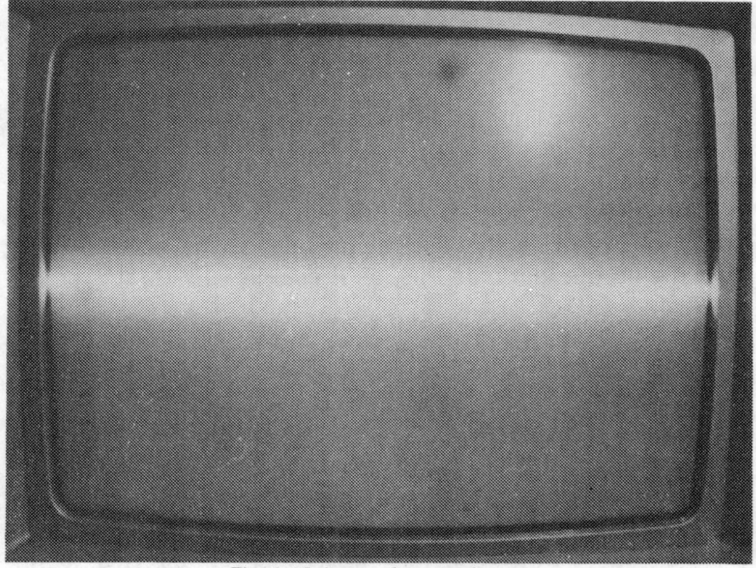

Fig. 1-12. Loss of vertical sweep.

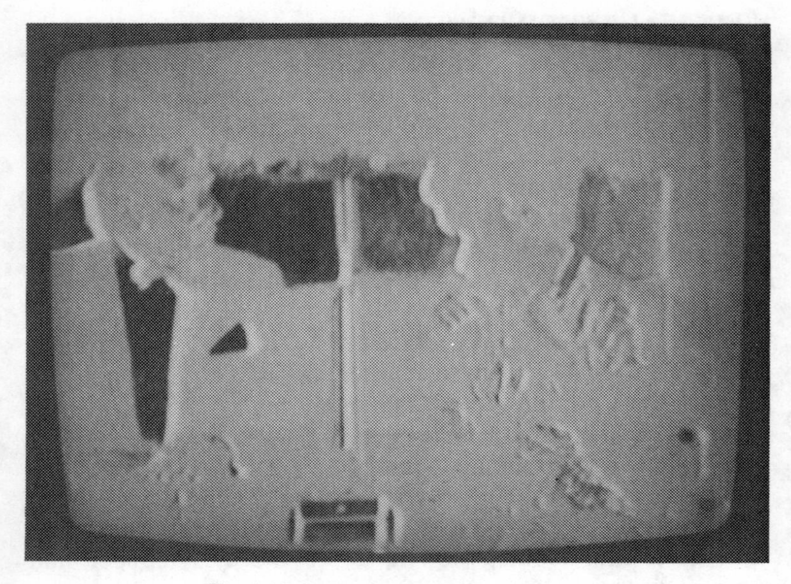

Fig. 1-13. Negative picture—video trouble.

probably good. You will then need to change out other modules or look for troubles in the main chassis. Be on the lookout for faults in the main chassis that may feed high voltages or pulses into the replacement modules, blowing them out also. In some cases two or more modules may be faulty in the set, as sometimes occurs when there has been some lightning damage.

When a new module is installed and the set's symptoms *change*—although the set may still not be operating correctly—then either the old module was defective or the new module might be faulty in a different way. In other words, all you know for sure is that the two modules are not alike in some way. Be on the alert for this, as it could really throw you off track.

At times you will actually have to troubleshoot the plug-in modules with the scope and VTVM in order to locate problems in other areas of the chassis. Let's use a GE MA color chassis that uses plug-in modules as an example. This set produced a negative black-and-white picture as shown in Fig. 1-13, which indicated trouble in the video module. A new video module was installed but no change was noted. Just to be sure, another video module was pulled out of a working set and tried, again with the same result. This, then, called for some checks in and around the video low-level module. Now for a quick look to see what was found with this negative-picture set.

Refer to the partial schematic of the video module in Fig. 1-14. A scope was used to look at the video signal at pin P4 of the module (also the base of the transistor), and it appeared normal at this point. However, at the collector of the video driver transistor the signal had the same polarity and a little less gain than at the base. The signal should have been inverted by this stage and also amplified, so this stage was evidently the troublemaker.

A reading taken at the collector with a VTVM indicated a zero voltage. It was found that the +23 volts at pin P3 was missing due to an open PC connection on the main chassis. With no collector voltage the stage could not amplify or invert, so it acted as a resistance to pass the video signal on through to the other stages in an uninverted condition. Thus a negative picture resulted.

SUMMARY

You can now put to use some of the points found in this simplified, picture diagnosis introduction when using the following chapters of this book.

Find the chapter for the brand of set you are having a problem with, then look for the picture symptoms. A brief

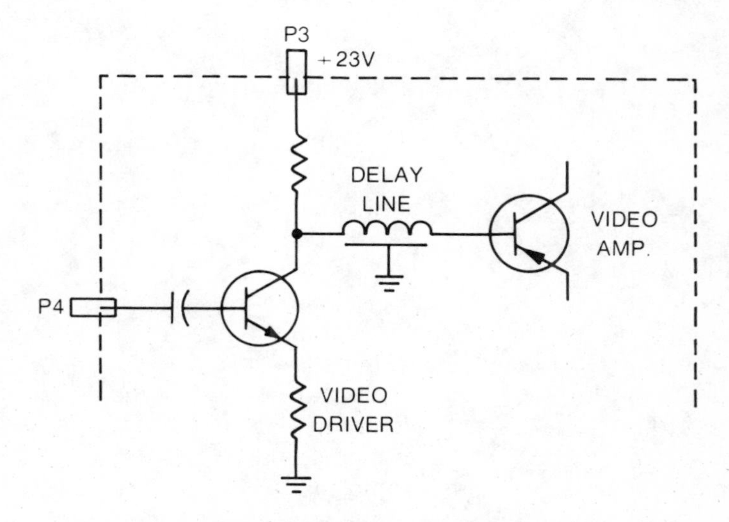

Fig. 1-14. Video module, partial schematic.

picture symptom is given, followed by the circuit probably at fault. For a wrap-up of each photo-symptom, the actual faulty component is called out, along with any other possibilities to check out. As you can see, then, this book can be used to learn as you earn.

The sets described in this book are all of American manufacture and include Admiral, General Electric, Magnavox, RCA, Sylvania, and Zenith. Other sets with similar problems may exhibit slightly different symptoms, so be sure to check out the other sections as well.

Weak or Snowy Picture

It is quite natural to suspect the RF input or tuner circuits when the picture is snowy and has that characteristic graininess. This indicates the set is either not getting enough RF signal or there is a fault within the set. Your job is to find that fault. The antenna and lead wires should be checked, as well as the cable system if you area has one. A test TV set can be used to confirm the signal conditions.

Should everything check and look correct, then the case histories found in this book will prove their worth. However, the picture symptoms shown in this book are not all-inclusive. Some may not be common to you, but deserve to be documented. And the cures cannot be considered exhaustive. The problems discussed here have all occurred after the point where the antenna lead connects to the tuner input circuit. Do not overlook the high-pass interference filters or balun coils.

It is most helpful to keep an open mind when tracking down RF troubles (or troubles related to insufficient RF). The problem could be in the AGC, AGC delay, and video circuits. Try to let your troubleshooting be logical and systematic.

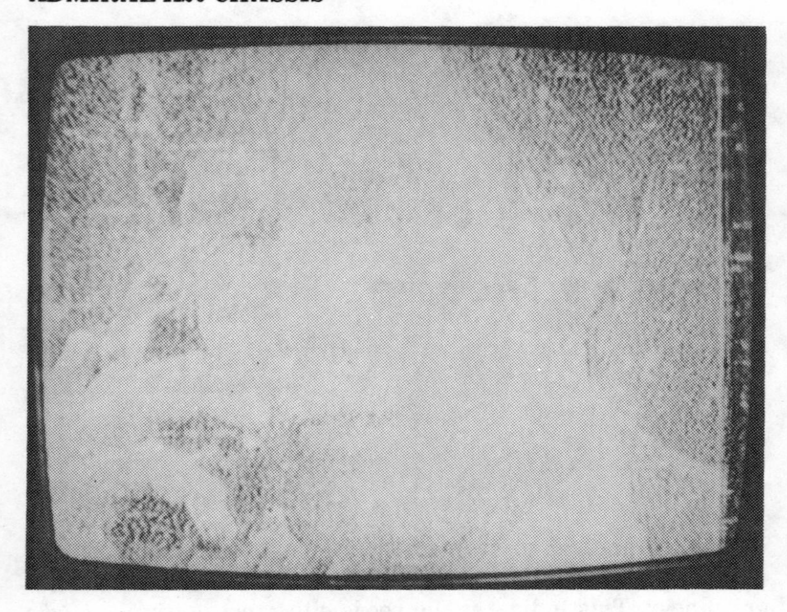

Picture Symptom: Picture is weak and snowy even on a strong station. Look for trouble in the VHF tuner or AGC delay circuit.

Circuit Operation: The delay circuit is used to prevent AGC action until a predetermined input signal level is reached. This sytem maintains full tuner amplification for moderate to very weak signals in order to achieve the highest signal-to-noise ratio. The delayed AGC is applied to the RF amplifier in the VHF tuner. The RF, like the IF AGC in this chassis, is a "forward" AGC design, since an increasing positive bias voltage decreases gain. The delayed AGC circuit is shown in Fig. 2-1.

A PNP transistor is used in this delayed AGC system. The positive voltage applied to the emitter is adjusted by the *AGC delay* control. The source of positive voltage for the base of Q506 is from the first-IF collector circuit and B+. The *AGC delay* control is adjusted to just reverse-bias Q506 on a weak signal. With this condition, as stronger signals are received, the forward IF AGC causes the first-IF transistor to increase conduction, which decreases its collector voltage. The increased voltage drop across R307 reduces the voltage to the base of Q506. The base voltage is then less positive with respect to the emmiter, and Q506 conducts. The collector electron-current path of Q506 is through resistor R4, the base biasing resistor for VHF amplifier transistor Q1. The voltage produced by this current makes the base of the NPN amplifier more positive, which then increases its conduction to produce the desired forward AGC. The RF AGC voltage developed depends on the voltage charge at the base of delay transistor Q506.

Probable Cause: This snowy picture was caused by very high leakage in C512, a bypass capacitor on the tuner AGC line. A defective Q506 (AGC delay transistor) or an open *AGC delay* control could cause the same symptoms.

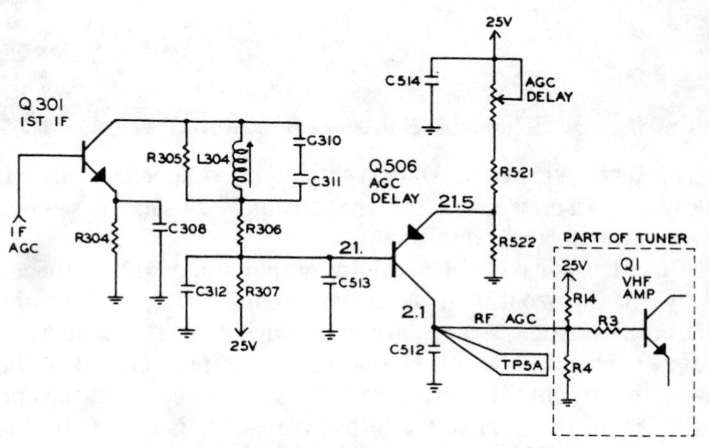

Fig. 2-1. Admiral K20 chassis, delayed AGC circuit.

Picture Symptom: Dark vertical lines appeared on left side of screen on weak or inactive channels. In some cases this was an intermittent condition.

Circuit Operation: Horizontal output stage Q103 is turned on by the horizontal drive pulse. Refer to Fig. 2-2 for the horizontal output sweep system. During Q103 conduction, heavy current flow in the sweep windings provides the sawtooth current for the horizontal deflection coils. At the end of line scan, the horizontal output transistor is cut off by the drive current, which allows the sweep field to collapse. This effect causes a large positive pulse to be produced in the sweep transformer.

A horizontal pulse from the flyback at terminal 17 is a source for keying and blanking functions. At the time the pulse returns to zero, damper diode conduction takes place in D101, restricting the pulse from going negative and thus preventing unwanted oscillations or ringing.

Capacitor C107 is used for yoke tuning. L101 and C106, connected between terminals 13 and 15 of the sweep transformer, tune the system to the fifth harmonic of the horizontal sweep rate. Horizontal yoke current at terminal 3 is coupled to the horizontal coils via the pincushion transformer.

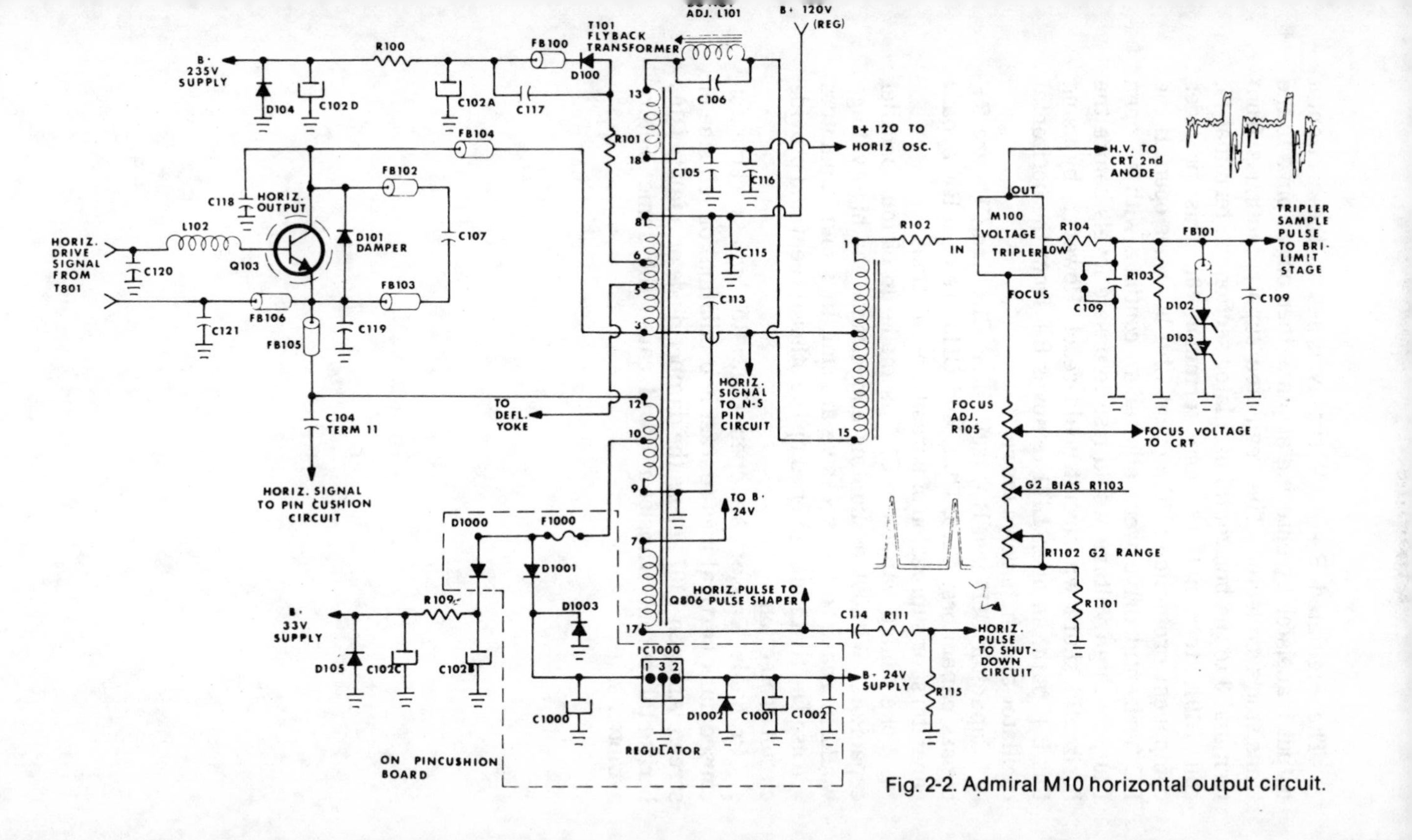

Fig. 2-2. Admiral M10 horizontal output circuit.

The regulated B+ 120V supply is fed to the horizontal output transistor by entering at pin 8, then out at pin 3 to the transistor's collector. The regulated supply continues from terminal 3 to the horizontal oscillator section via terminal 15, coil L101, terminal 13, and terminal 18. This provides component protection in the event L101 becomes open. If the DC horizontal drive were allowed to continue with an open L101, the high voltage would rise to unsafe levels before the horizontal shutdown circuit could react. However, by using this B+ path, an open L101 removes B+ from the horizontal oscillator and shuts down the set.

Capacitor C115 is a B+ 120V filter. C105 and C116 are RF bypass capacitors. C119, C121, and C113 are also RF bypass capacitors for supression of parasitic oscillations.

Yoke tuning capacitor C107 is made up of four parallel capacitors. Without a shunting capacitance the high voltage would increase to an excessive amount, thus four capacitors are used so that if one opened up the other three would provide circuit protection.

Probable Cause: An open C113 RF bypass capacitor caused dark vertical ines (parasitic oscillation) on left side of screen. An open C107 across the damper diode or a faulty Q103 horizontal output transistor could cause the same picture symptoms.

ZENITH 15Y6C15 CHASSIS

Picture Symptom: This set produced a very poor, grainy picture. Another symptom could be a very weak, washed out picture. Look for trouble in the IF subchassis or AGC systems.

Circuit Operation: The IF amplifier system has three transistor stages as shown in Fig. 2-3. All are NPN types with the first-stage AGC controlled with "forward" AGC voltage applied to its base. An increase in positive voltage at its base will increase the collector current and decrease the collector-to-emitter voltage, reducing the gain of the stage.

For maximum gain the AGC bias voltage at the base of TR1 would be +4.5V while minimum gain would occur at about +7.0 volts. This voltage level is held contant by a clamp network in the AGC circuit.

Transistor TR1 is coupled to TR2 by double-tuned interstage transformer L5 in the collector circuit of TR1. The unique design of this circuit boosts the sound carrier at weak signal levels. TR2 is coupled to TR3 through a pi coupling network (L6, C25, and C28) to provide the proper Q and low-pass characteristics.

Coils L7 (primary) and L12 (secondary) form a slightly overcoupled, double-tuned circuit with a 41.25 MHz output trap in the coupling network between L7 and L12. Resistor R23 and

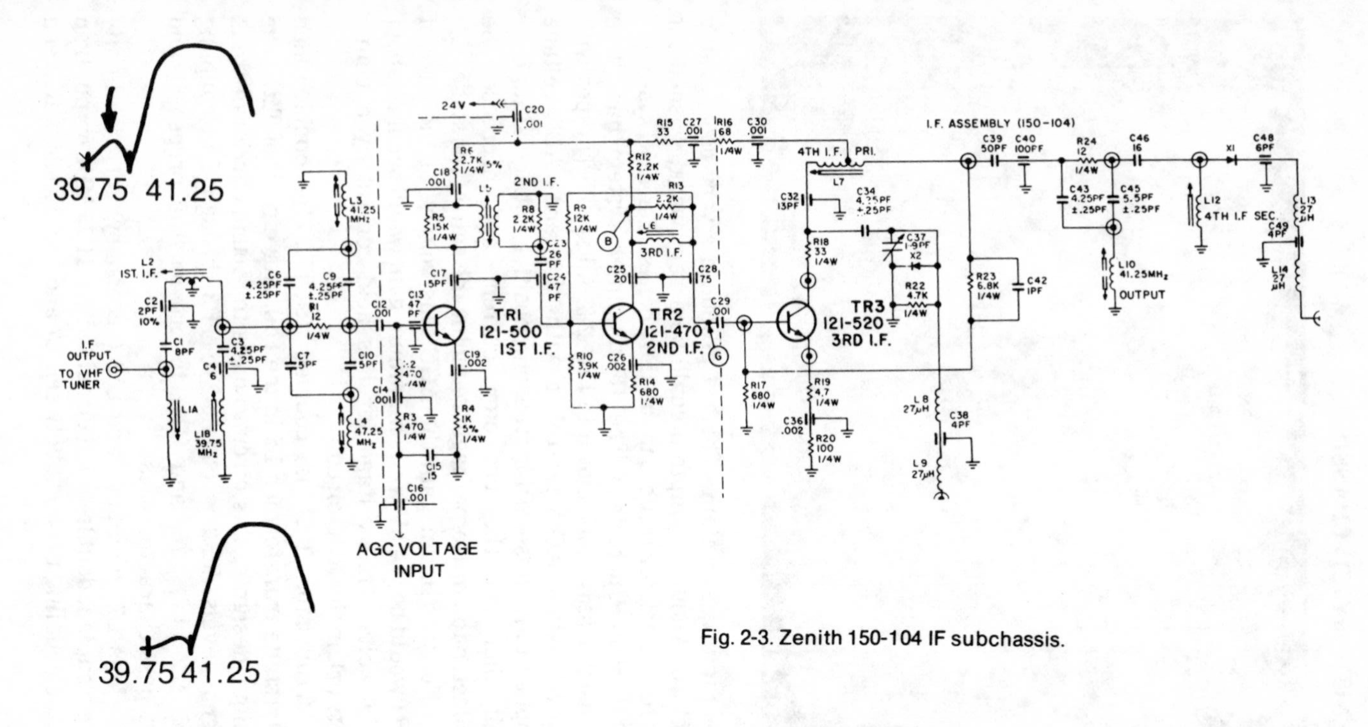

Fig. 2-3. Zenith 150-104 IF subchassis.

capacitor C42 form a partial neutralizing and DC stabilizing circuit for the base of TR3.

Detectors X1 and X2 (not shown) provide detection for the Y amplifiers and for sync-sound-AGC amplification.

There should be approximately a 4-volt peak falling to zero at test point C1 (video signal) and about 2-volt peak falling to zero at test point C2 (sound sync signal).

Probable Cause: Check for correct AGC and B+ voltage to the IF subchassis before troubleshooting the IF circuits. Next, check all DC voltage around the transistor stages.

This picture symptom was due to a decrease in value of R4 in the emitter circuit of TR1. For a weak, washed-out picture, check for a faulty (open) TR3, the third-IF amplifier transistor.

Signal tracing or injection can also be used to isolate problems in the IF stages.

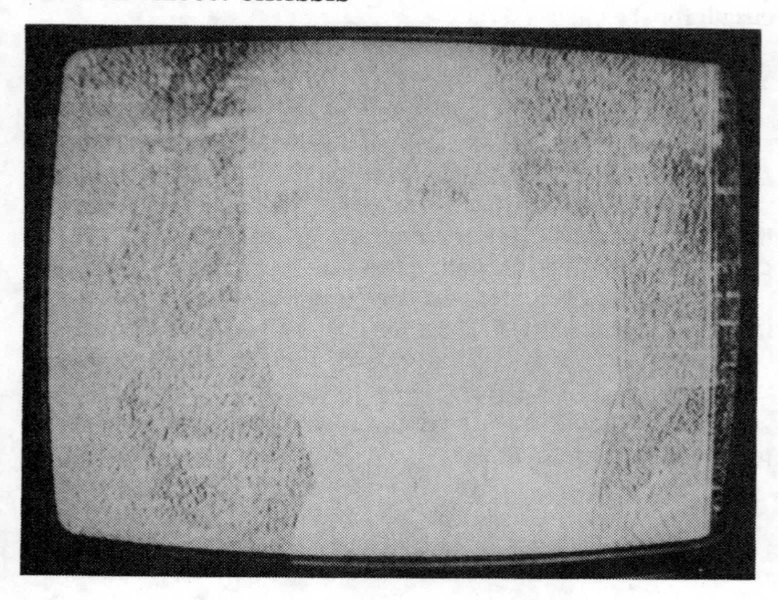

Picture Symptom: Drive lines appear at right side of raster. Generally Barkhausen, "snivets," "spooks," "jail bars," and drive lines are all related types of interference and are caused by high-frequency oscillations (usually within a tube) that radiate from improper lead dress as well as from horizontal stages with faulty components.

Circuit Operation: Refer to the horizontal sweep circuit in Fig. 2-4. The damper tube is responsible for tracing the raster line from left to center. The horizontal output tube is used for tracing from center to right of the screen.

With this information, drive lines, snivets, spooks, and jail bars can be analyzed to help isolate the faulty stage. If these drive lines occur on the left side of the raster, suspect the damper circuitry; if on the right half the horizontal output tube circuitry would be at fault.

Drive lines that appear as narrow, bright, vertical lines in the raster are most always associated with the horizontal sweep circuitry. Excessive cathode current through the tube, which may be caused by an improper control grid waveform, will contribute to this fault. In rare instances drive lines can be caused by excessive ringing current in the damper stage and should not be overlooked.

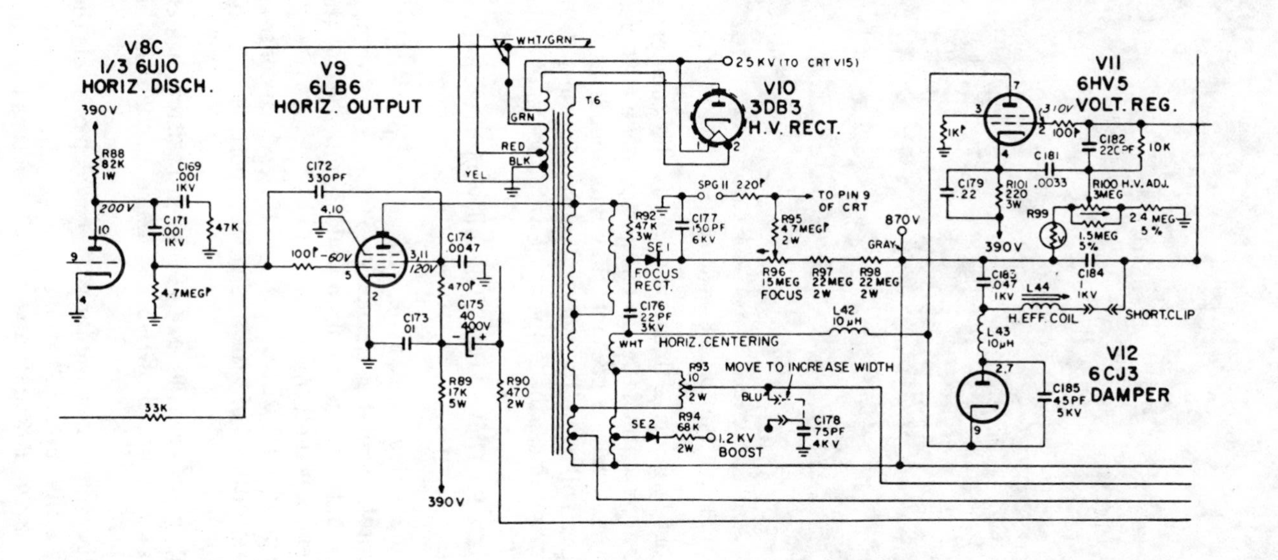

Fig. 2-4. Zenith 14Z8C50 chassis horizontal sweep circuit.

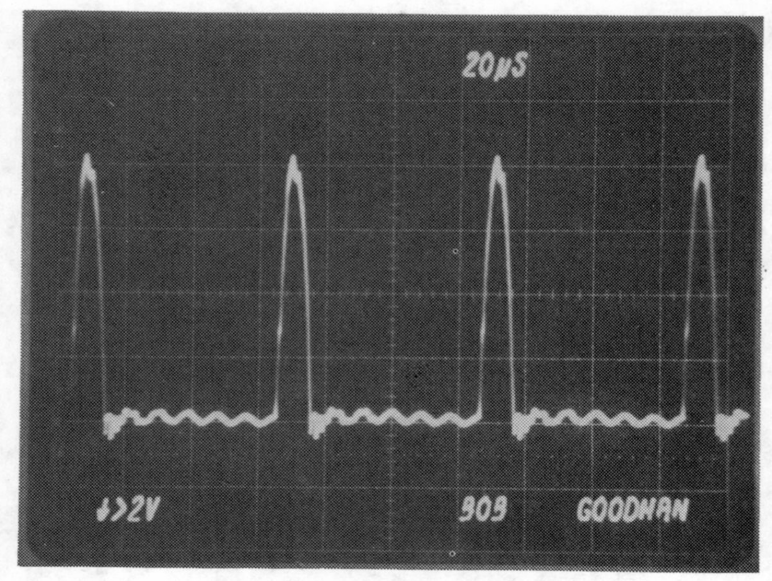

Fig. 2-5. Horizontal blanking pulse.

To check out faint drive lines, place an oscilloscope probe close to the damper cathode lead—but do not make electrical contact. If drive lines are present, they will appear as very small notches or spikes on the pattern between pulses. If the set has some drive lines, the scope trace will appear like that shown in Fig. 2-5.

Service Hints for Drive Lines: The following are common causes of drive lines:

- A wrong value of resistor R88 in the plate of the horizontal oscillator discharge stage. The resistor value controls the peak-to-peak amplitude of the trapezoidal waveform.
- A defective 0.001 μF capacitor (C169), which controls the amplitude of the sawtooth portion of the waveform.
- Wrong value of the 47K peaking resistor, which controls the amplitude of the pulse portion of the waveform.
- A faulty coupling capacitor (C171), which controls the overall amplitude of the waveform.
- Coils L42 and L43 and capacitor C185 across the plate and cathode of the damper play an important part in

eliminating a "spook" type of interference, which causes "blobs" on the left side of the raster. The damper tube, which supresses the primary ringing of the sweep transformer, can cause a spook during the fast transistions of current through the tube (an oscillatory effect). The two coils present a high impedance to block this effect, and the capacitor provides a short circuit of the radiating elements (plate and cathode). These components should be checked if such an interference pattern is exhibited. Also, the value of C185 controls the width of the raster, and if this value is too high, excessive width may result.

Probable Cause: This drive-line symptom was caused by a faulty C169 capacitor.

ZENITH 12B14C52 CHASSIS

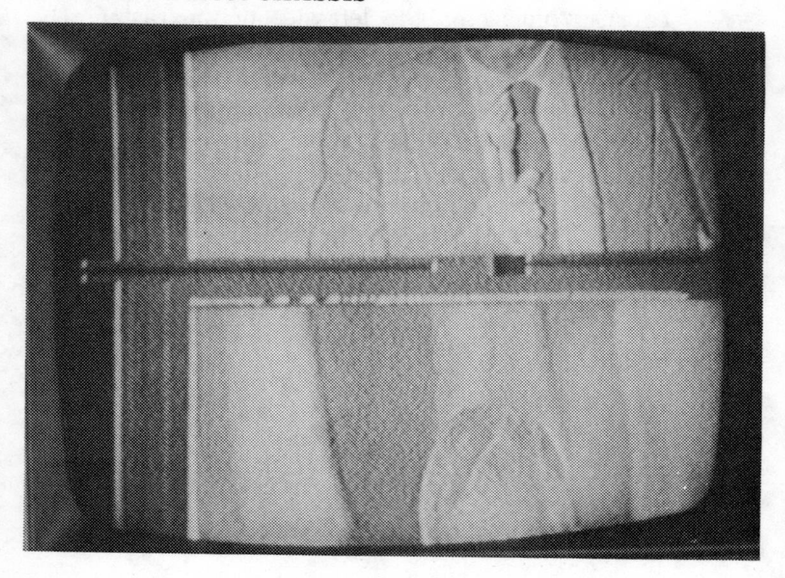

Picture Symptom: Picture has no vertical or horizontal lock. Picture is also weak and washed out, and appears to have AGC or sync circuit trouble. (Look for this trouble in most color chassis in the sync/AGC control system; however, this set has an added noise amplifier circuit that caused the trouble, so let's take a look at this noise amplifier.)

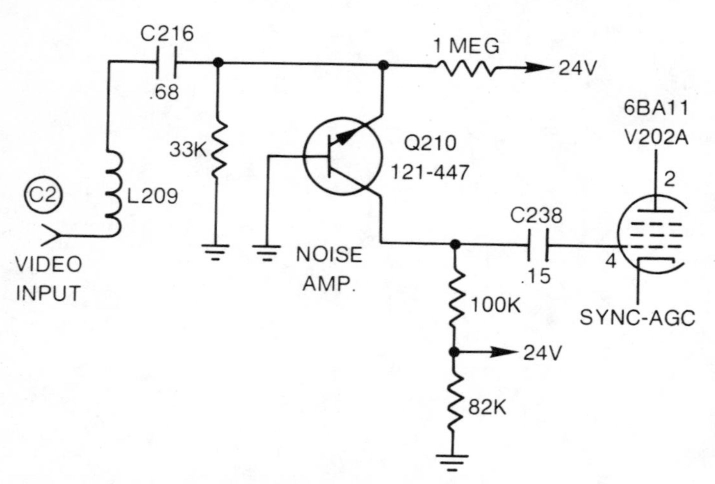

Fig. 2-6. Zenith 12B14C52 chassis, noise amplifier.

Circuit Operation: Refer to the simplified noise amplifier stage in Fig. 2-6 that has been incorporated in this chassis. This is a transistor stage that is being operated as a grounded-base amplifier. It is biased just beyond sync tips so that only noise pulses are amplified from test point C2 (video information) before being coupled to the control grid of the AGC/sync amplifier stage. In this manner, the noise signal is sufficiently amplified to provide for an improved noise-gating action.

The composite video signal at test point C2 is coupled through coil L209, through capacitor C216, and to the emitter of Q210. Biasing networks for the emitter and collector are provided by a 1M and 33K resistor in the emitter circuit and a 100K resistor from 24 volts in the collector circuit. The output from the collector (zero phase reversal) is coupled to the AGC control grid of V202A via coupling capacitor C238, a 0.15 μF.

Probable Cause: This loss of picture sync resulted from a defective (shorted) noise amplified transistor, Q210. An open Q210 would not have much affect on the picture reception, but you might notice that the sync system would not have any noise immunity. A scope can be used to check the signal through this circuit to see if Q210 will amplify.

A quick check for cause of sync trouble in this circuit is to pull out the noise amplifier transistor. If the picture now locks in, you know the trouble is around Q210. This circuit is used in only a few Zenith chassis, so be on the alert for this one.

Picture Symptoms: The set had to be constantly re-fine-tuned to have good color and a clear picture. The trouble appeared to be local-oscillator drift in the VHF varactor tuner but a new tuner had the same drift problem. So, the AFC module was suspected. This tuner drift could be caused by a faulty varactor tuner, or AFC module, or trouble in the nerve center module that supplies the DC tuning voltages for the tuner.

Circuit Operation: With the use of a varactor tuner, the AFC system operation is somewhat different than the AFC operation in the previous Zenith chassis. When checking the AFC circuit in Fig. 2-7, you will note that the input circuit has been changed to a tapped inductance instead of a capacitively coupled circuit. This provides an improved impedance match to the base of the amplifier transistor. The primary difference, however, is in the output circuitry.

The previously used AFC modules had a single output. Under conditions of no correction (or AFC defeat) the output voltage was +3 volts. When correction was required, the voltage produced by the discriminator circuitry would either add to or subtract from the nominal voltage.

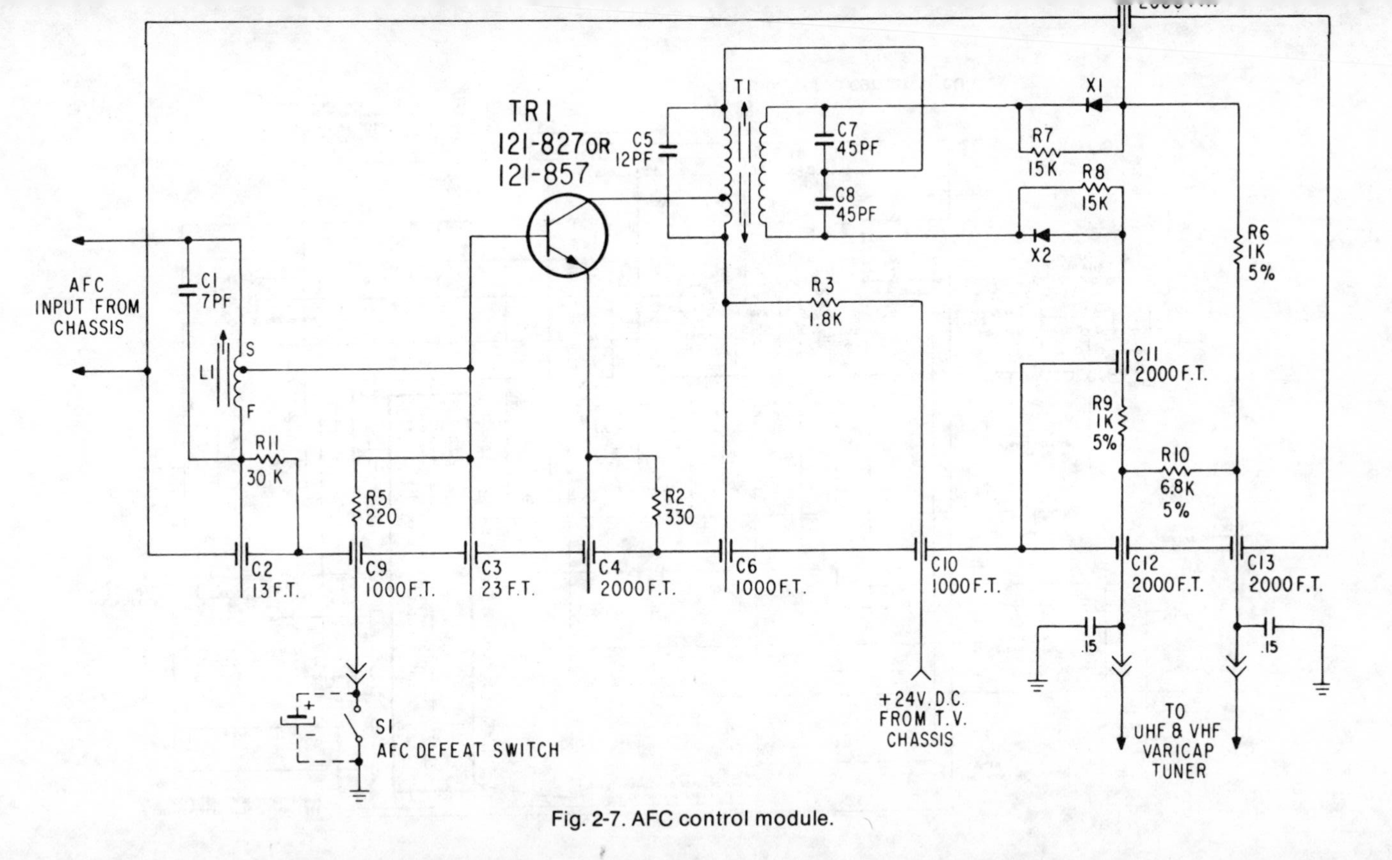

Fig. 2-7. AFC control module.

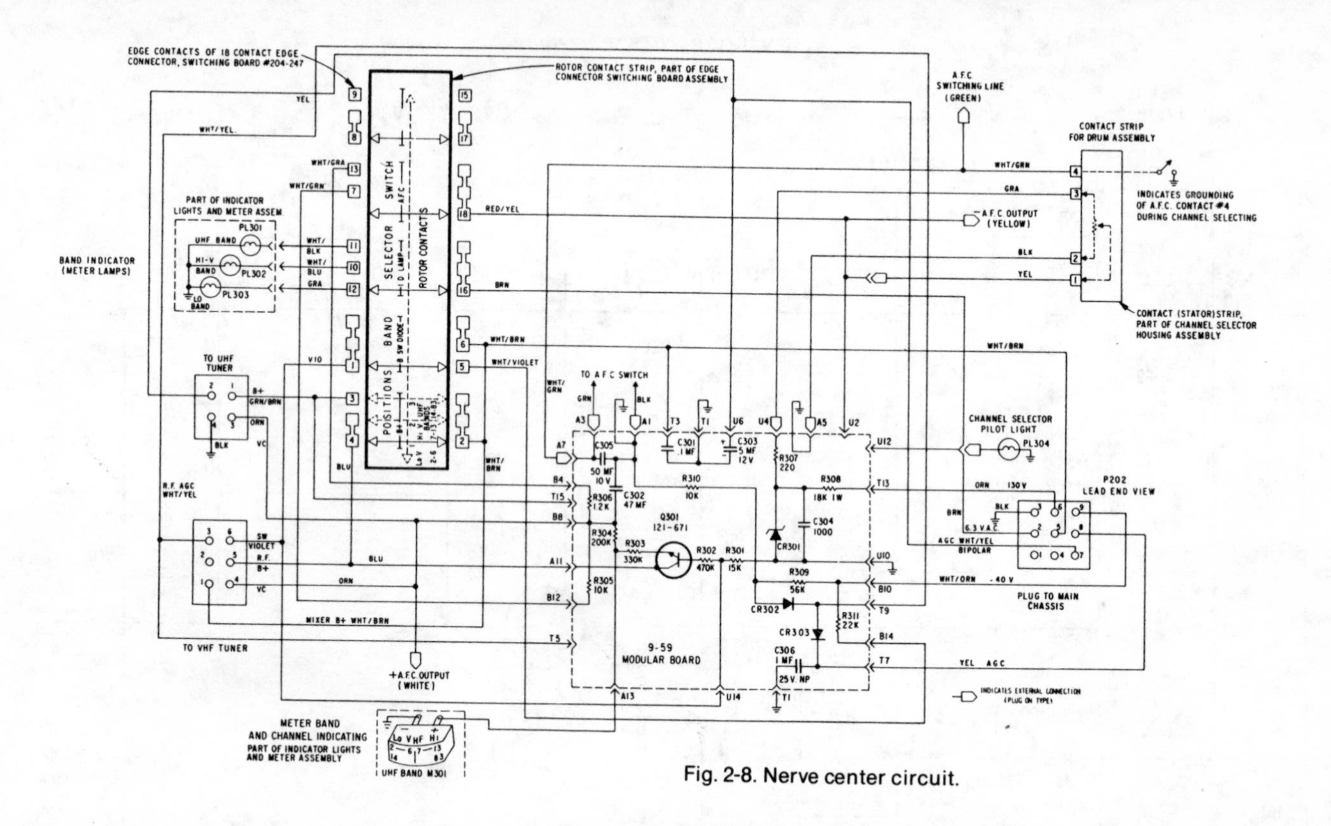

Fig. 2-8. Nerve center circuit.

The new AFC module uses a dual output. Under conditions of no correction (or AFC defeat) the voltage differential between the two outputs is zero. As correction is required, the two outputs become unbalanced and produce a voltage differential as high as ±1 volt (on low VHF range) across resistor R10.

The varactor tuning voltage, from the 130-volt B+ supply in the main chassis, is connected to the nerve center (Fig. 2-8). This +130 volts is coupled through R308 to zener diode CR301. The resultant +33 volts is connected to the drum assembly, stator contact strip, contact 3. As the drum is rotated to each position, the total resistance of each tuning resistor is placed across stator contacts 2 and 3. The slider (center arm) is connected to stator contact.

The selected tuning voltage, available at contact 1, is connected to the AFC module at feedthrough C13. The voltage, coupled through R10 and available at the other output, is connected to the varactor tuning-voltage inputs at both tuners. Again, under conditions of no correction, there will be no voltage drop across R10, and the varactor tuning voltage will remain as set by the tuning resistor. As correction is required, the voltage drop across R10 (positive or negative with respect to the tuning voltage) will either add or subtract from the preselected tuning voltage.

On the VHF low band, the varactor tuning voltage is coupled directly through the AFC module to the VHF tuner.

On the VHF high band, resistor R305 (10K mounted on nerve center) is placed in parallel with R10 in the AFC module. This tends to limit the AFC voltage swing, thus maintaining the same frequency pull-in range as the low-band function.

On UHF, resistor R305 is switched out and R306. (1.2K) is placed in parallel with R10. The decreased resistance serves to further limit the AFC voltage swing since less voltage correction is required at the higher frequencies. Stator contact 4, in parallel with the *AFC defeat* switch, is grounded during drum rotation. This prevents AFC lock-up on erroneous carriers during channel change.

Probable Cause; This channel drift problem was due to leakage in diode X1, which appeared to develop more leakage as the receiver warmed up. Faulty varactor diodes in the tuner or unstable DC tuning voltages could cause the same symptoms.

Video Defects

When the chassis has power and audio but no video at the CRT, there are many possible faults that could have occurred—horizontal sweep, yoke, HV rectifier or tripler, video amplifier and AGC systems faults. The appearance of a raster greatly simplifies matters, for it lets you rule out problems in the horizontal sweep, deflection, and CRT high voltage supply. Other video troubles are smeared and intermittent video.

Where there is no raster at all, you may find other clues that will help narrow down the list of possibilities. Can you obtain a line across the screen with switch in the *setup* position? If so, look for a problem in the video amplifier. Is there a trace of hum in the audio? Check for excessive current drain in the horizontal sweep output tube (cherry red plates). If the plates are glowing red, look for trouble in the horizontal oscillator, as the sweep output tube may not be receiving enough drive signal. There could also be a short loading down the sweep output system. In these self-biased circuits, big trouble can develop, for the unbiased horizontal output stage draws more and more current until the tube (at the very least) is destroyed, and very possibly many other components in the sweep output stages. Also be on the lookout for holddown and feedback regulation circuits in the control grid of the horizontal output tube stage.

With no horizontal drive to a transistor horizontal sweep output stage, the system will usually just shut down, and no damage will occur.

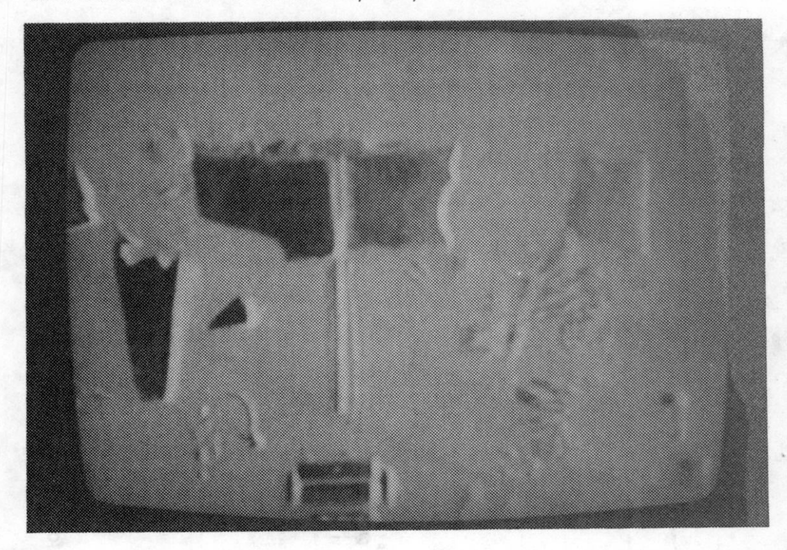

Picture Symptom: This set produced a negative black-and-white picture. The color picture information was normal. Substituting new modules did not solve the problem. The video low-level module was not at fault, but an external condition made it appear that it was.

Circuit Operation: The video low-level module (Fig. 3-1) is used for the following operations:

- Video reference shift.
- CRT beam limiting.
- Impedance matching to the delay line.
- Low-level video amplification.

Isolation from the video-IF module, along with impedance matching to the input of delay line DL301 is accomplished by class-A amplifier Q301. Note that the video signal is inverted in this stage. The output of emitter follower Q302 drives the base of first video amplifier Q303 through contrast control R1904. Y301 and Y302 clamp the video to a DC level determined by the amplitude of the composite sync. The signal is further amplified by Q304, the reference shift amplifier. In this stage, negative horizontal and vertical pulse are introduced.

The CRT beam limiter establishes a maximum limit for the CRT beam current so that during the worst case of

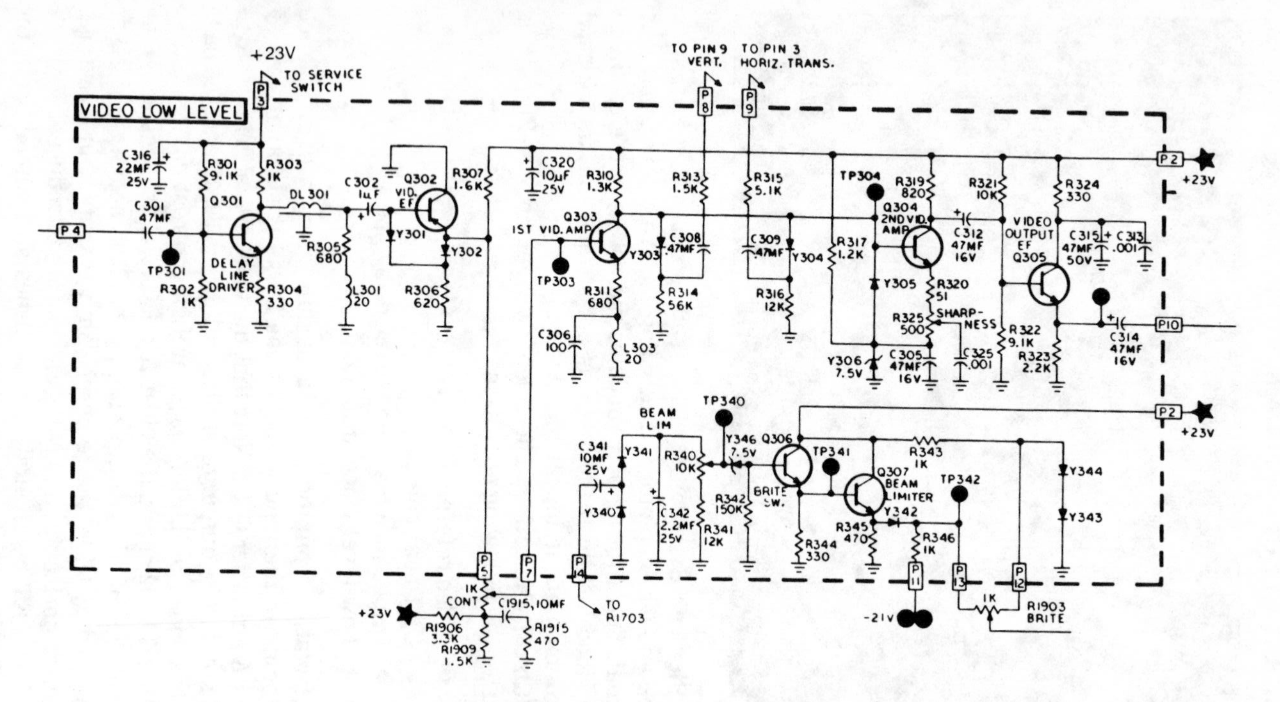

Fig. 3-1. Video low-level module for GE MA chassis.

customer control misadjustment, the rating of the CRT cannot be exceeded.

Zener diode Y346 is used as a threshold device. *Beam limiter* control R340 is adjusted so that zener Y346 conducts when the CRT beam current reaches 1.7 mA. Q306, acting as an emitter follower, provides a very high input load impedance to prevent the voltage developed across C342 from being affected by any loading after zener conduction is reached.

Probable Cause: The negative picture was caused by a loss of the +23 volts at pin 3 of the video low-level module. A break in the circuit foil in the main chassis to pin 3 caused loss of voltage. Most of the schematics do not show the +23 volts at this point, and this fact can easily be overlooked. With no voltage at the collector of Q301, the stage will not amplify or invert the video signal, thus the negative picture will appear and the stage is seen as a low resistance to the video signal.

Picture Symptom: Very dim or dark picture with no video information. The only sound coming from the speaker was a low whistling or warbling noise.

Circuit Operation: Several areas of the receiver could be at fault. The defect could be in the tuner, IF stages, video amplifiers, or AGC system. Because this was a video fault, a look at the video circuit will be in order.

The video signal circuit (Fig. 3-2) consists of two transistor amplifier stages and a 6AG9 pentode tube output stage. First video amplifier Q201 is a silicon NPN transistor connected as a phase splitter. At the emitter of Q201, the composite video signal is about 2 volts peak-to-peak. This signal is in phase with the base signal and is fed to the color section of the receiver.

A phase-reversed signal appears at the collector of Q201 with an amplitude about three times that of the base signal. At this point, composite video is picked off to be fed to the sync clipper and AGC stages. The collector of Q301 is connected directly to the base of Q202, the second video amplifier stage which is not shown. Q202 is a PNP silicon transistor connected in the common-emitter configuration but, because of the large amount of degeneration in the emitter circuit and the low

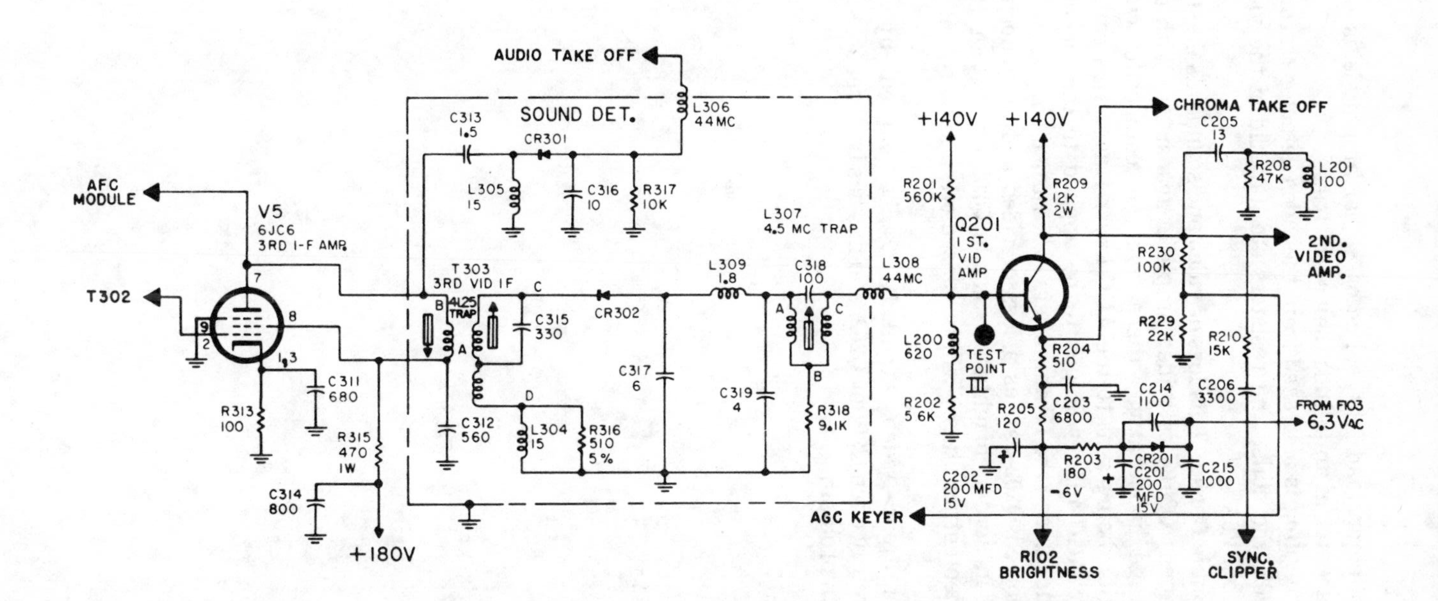

Fig. 3-2. Video circuit of GE KE chassis—third IF amplifier, sound detector, and first video amplifier.

value collector load, the collector signal amplitude is only slightly greater in amplitude than at the base.

DC coupling is also used from the video detector (through Q201, Q202, the coils, and resistors) to the grid of the video output tube, V6A.

Probable Cause: This loss of video and sound was due to a faulty (open) Q201 first video amplifier transistor. With this video stage inoperative, there was no composite video information being sent to the AGC keyer stage. This caused improper AGC bias voltages to be applied to the video-IF stages and caused the IF amplifiers to be turned off.

A shorted Q202 causes a loss of video and an abnormal increase in screen brightness. Should Q202 open, V6A (the 6AG9 video output stage) will be biased off, and there will be no raster.

Intermittent video is likely to be caused by an intermittently open Q201 or Q202, and this has been reported as a common problem.

GENERAL ELECTRIC C-2 CHASSIS

Picture Symptom: Dark area at top of picture; otherwise, good clear picture and sound. Look for trouble in the video amplifier and blanker circuits.

Circuit Operation: As shown in the Fig. 3-3, video-circuit blanker transistor Q305 switches the CRT off during the retrace interval. Sweep retrace pulses coupled to the base cause Q305 to conduct during retrace time. This has the effect of shorting the output tube grid to ground and cutting off the CRT beam current. Between retrace pulses, when video information is being transmitted, Q305 is effectively an open circuit and has no effect on the operation of the TV receiver. The signal at the collector of Q305 is a composite of the detected video signal and the internally generated vertical and horizontal retrace pulses. Note that direct coupling is used from the video detector, through Q301, Q304, and onto the grid of the video output tube, V5A.

Probable Cause: If blanker Q305 shorts it will kill the raster. An open Q305 can be detected by rolling the picture vertically, since if retrace lines appear, Q305 is not working.

A shorted Q304 will result in a loss of video and an abnormal increase in screen brightness. Should Q304 open, V5A will be biased off, and there will be no raster. Video that goes off and on is usually caused by an intermittent Q304 second video amplifier.

The dark top of the raster was caused by a fault (leakage) in the Q305 blanker transistor.

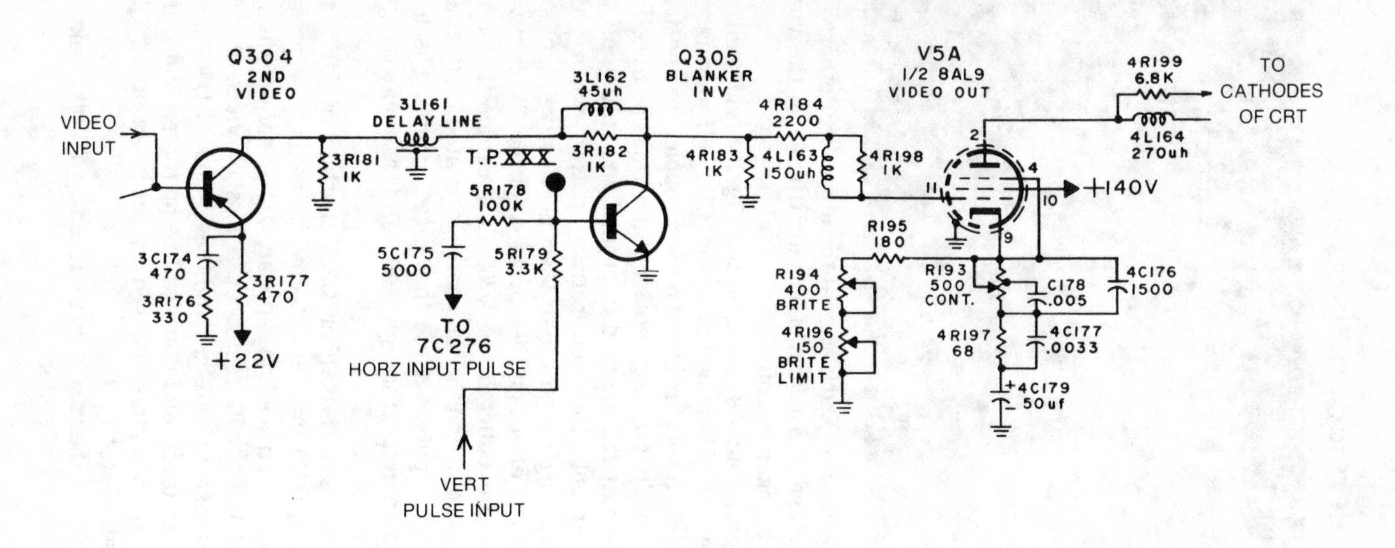

Fig. 3-3. GE C-2 chassis, video and blanker circuit.

GENERAL ELECTRIC C2/L2 CHASSIS (HYBRID)

Picture Symptom: Weak, washed-out picture with very little contrast. But good screen brightness, correct HV, and sound okay. Look for trouble in the video amplifier stages.

Circuit Operation: The video signal is recovered by diode 3Y153 and fed to the base of the first video transistor at test point 3. Referring to the schematic in Fig. 3-4, we see that the video signal circuit consists of two transistor amplifier stages and a tube-type video output stage that is not shown. First video amplifier Q301 is a silicon NPN transistor connected as a phase splitter. At the emitter of Q301 the composite video signal is about 2 volts peak-to-peak, which is very nearly equal in amplitude to and in phase with the signal at the base. The emitter is the source of chroma signals for the color bandpass amplifier section. A phase-reversed signal appears at the collector of Q301 with an amplitude about three times that of the base signal. The bases of Q304 and Q303 are connected directly to the collector of Q301. Q303, an emitter follower, links the video circuits to the AGC and sync circuits.

Transistor Q304, the second video amplifier, is connected in the common-emitter mode. The collector load consists of a 1000Ω resistor (3R181) in parallel with the delay line, a

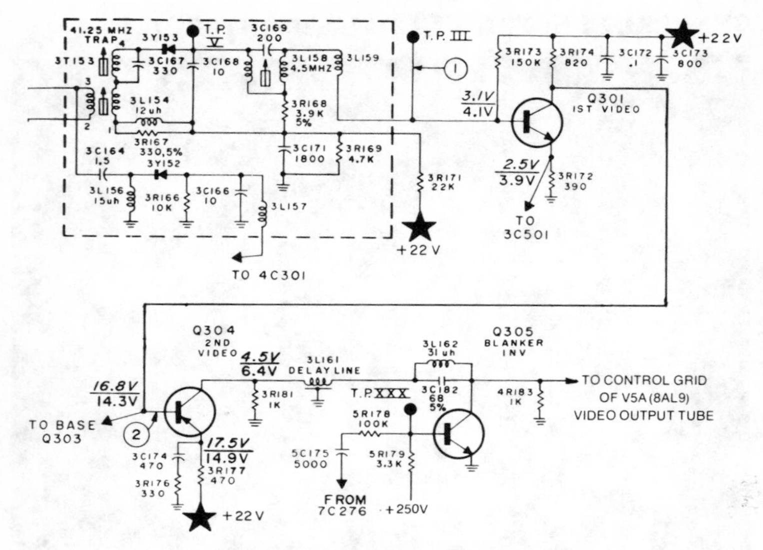

Fig. 3-4. GE L-2 chassis, video circuit.

peaking coil, and another 1000Ω resistor. The DC path through these components is also shared by blanker Q305 and the grid of video output tube V5A.

Because of the large amount of degeneration in the emitter circuit and the relatively low value collector load, the collector signal amplitude is only slightly more than the input signal amplitude on the base. Blanker Q305 switches the CRT off during the retrace interval. The signal at the collector of Q305 is a composite of the detected video signal and the internally generated vertical and horizontal retrace pulses. This signal goes on to the control grid of V5A. The plate of V5A is direct coupled to the three CRT cathodes. Thus, any changes in the DC current flow through the video output tube will cause a brightness change on the CRT screen.

Probable Cause: The direct-coupling circuit can be used to troubleshoot the video system. For a blank raster with no video, use the following checks to isolate circuit fault.

- Varying the brightness control setting should change the brightness level. If not, the defect is in the video output stage.
- Shorting one end of the delay line to ground should decrease the brightness level. If it does not, the fault is between this point and the video output stage.

- Shorting test point 3 to chassis ground should decrease the brightness. If not, the defect is between this point and the delay line.

Should blanker Q305 become shorted, the screen will be black. To check for an open Q305, roll the picture vertically, and if retrace lines appear, the blanker is not working.

A shorted Q304 will cause a loss of video and too much screen brightness. Should Q304 open up, V5A will be biased off, and no raster will be seen.

This particular symptom was caused by a faulty Q304 second video amplifier transistor that provided no amplification.

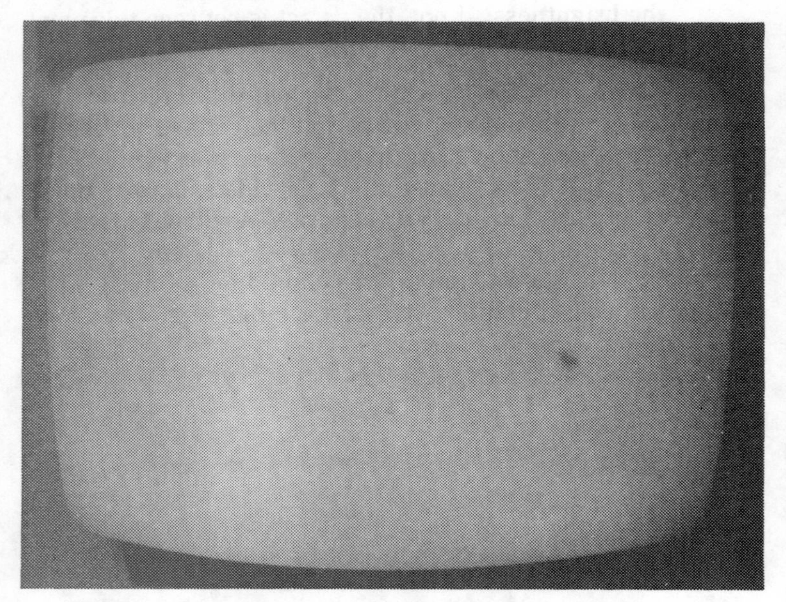

Picture Symptom: On a blank screen the picture has a lighter shade on the left-hand side. Look for trouble in the video amplifier section.

Circuit Operation: First video amplifier Q201 (Fig. 3-5) is an NPN transistor connected as a grounded emitter. The base bias is derived from the voltage at the diode load resistor. Since this voltage is about −1.5V, Q201 would not conduct if the emitter was grounded. To make Q201 conduct, a negative bias of about −2.0V is fed to the emitter. This bias is produced by rectifying the 6.3V AC heater voltage. The bias supply consists of the rectifier CR210 and the filter network. The output of the supply is −6.0V DC with a drop across the three resistors producing −2.0V at the emitter. The impedance is sufficiently high so that the reflected impedance at the base does not upset the alignment of the 4.5 MHz trap. This impedance also provides good phase response at the chroma take-off capacitor C703, so that there is no distortion of the chroma signal as it is fed to the chroma amplifier.

The amplified signal at the collector of Q201 is further amplified by second video amplifier Q202 and third video amplifier V6A. The second and third stages are conventional video amplifier circuits.

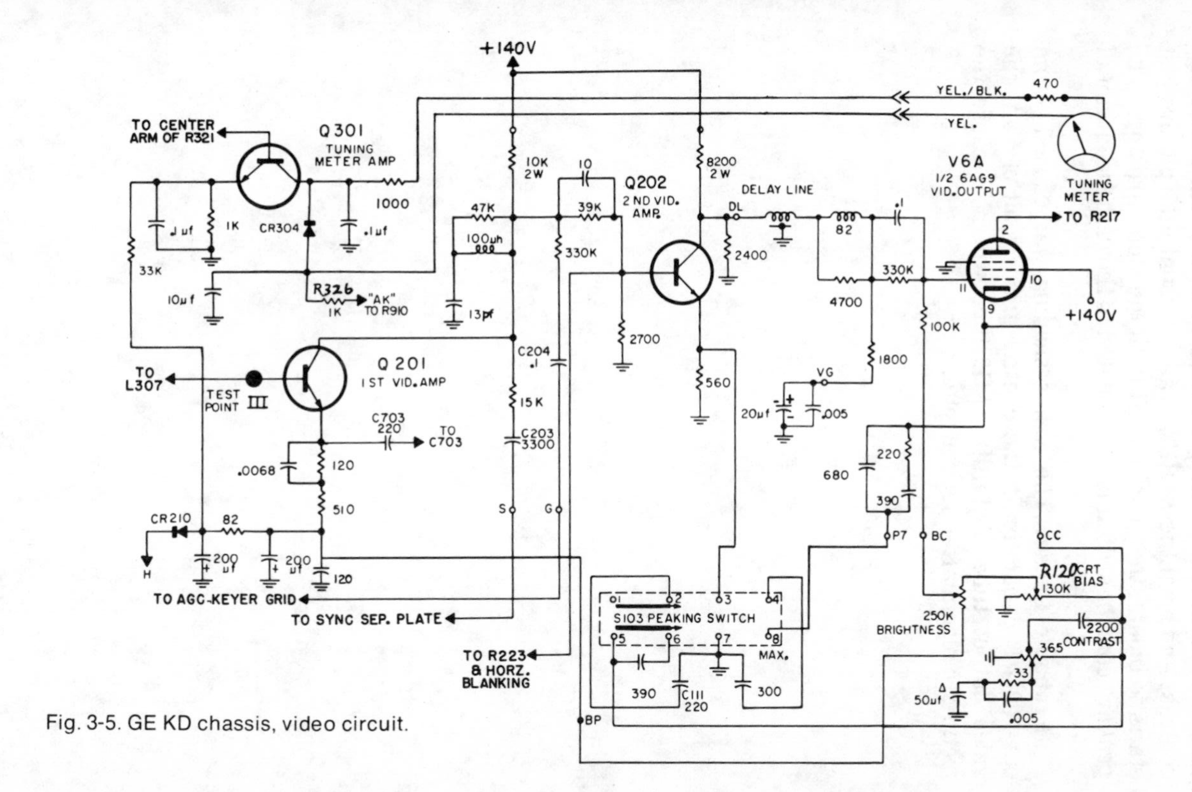

Fig. 3-5. GE KD chassis, video circuit.

The cathode circuit of V6A contains the contrast control and CRT bias pot R120. Bias from the −6.0V bias supply is fed through brightness control R120 to ground. With maximum contrast and brightness, R120 is adjusted until the picture just starts to bloom, then backing off until the blooming stops. The *peaking control* switch is also located in the cathode circuit of V6A.

Probable Cause: The raster shading was caused by an open 20 μF capacitor in the control grid circuit of V6A, the video output tube. A faulty 6AG9 tube will also cause the picture to bloom after set warmup.

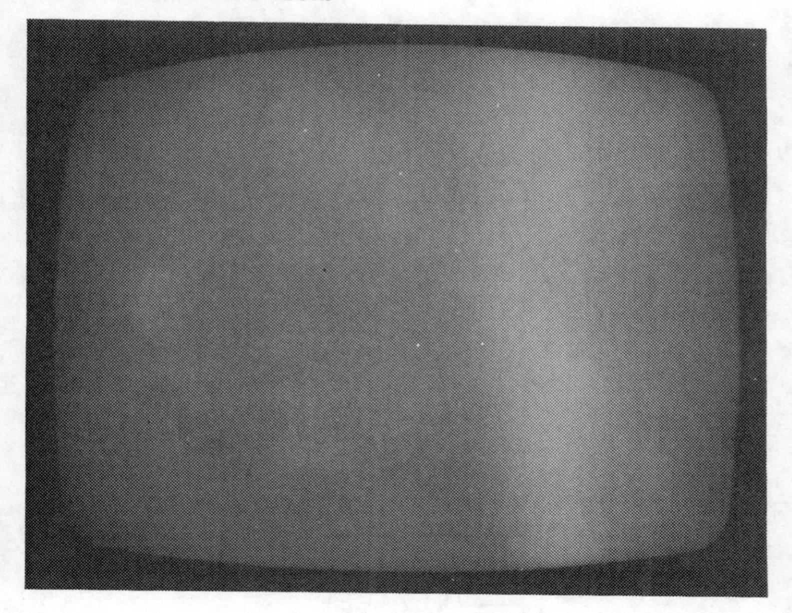

Picture Symptom: The side, white vertical bar on the right side of the screen was due to a horizontal blanking malfunction. Look for trouble in the horizontal blanking system.

Circuit Operation: Blanking pulse energy developed by the horizontal sweep output transformer (Fig. 3-6) is placed across a voltage divider network that consists of resistors R454, R456, and R458. The voltage developed across R458 is used as a drive pulse for blanking amplifier Q912. Blanking transistor amplifier Q912 is an emitter follower coupled to the emitter of fourth video amplifier Q908 by gating diode SC930. This diode is reverse biased by the voltage drop across emitter resistor R937 during trace time. However, during the retrace time, the pulse from R458 turns on Q912, developing a positive pulse in its emitter circuit. This pulse turns on diode SC930 and feeds the pulse to the emitter of Q908.

The fourth video amplifier's collector follows the emitter signal and applies a positive pulse to video driver Q910. This noninverting amplifier now feeds the positive pulse to the R, G, and B output amplifier emitters. The pulse-driven emitters maintain the correct pulse polarity in the collectors and blanks

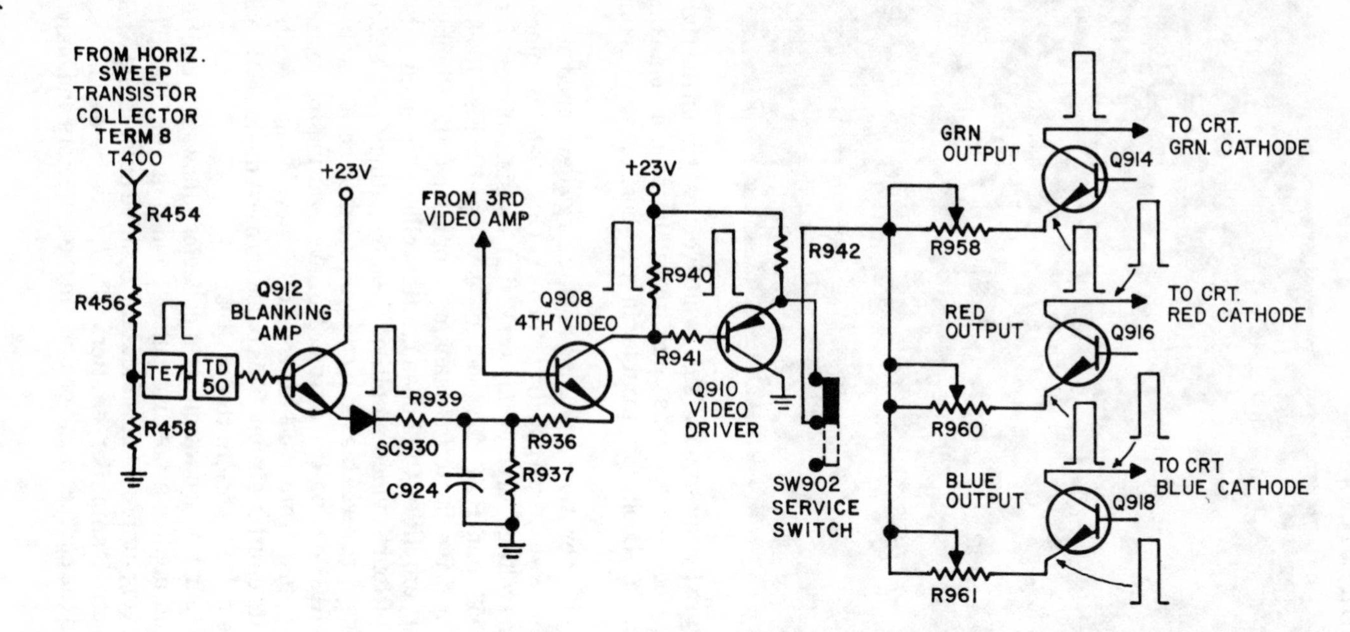

Fig. 3-6. Sylvania E05 chassis horizontal blanking circuit.

the CRT by driving the CRT cathode more positive than the grids during the horizontal retrace time. This action places the CRT into the black level just before trace time during each line. Other line irregularities are blanked out also.

Probable Cause: This loss of horizontal blanking was caused by an open R458 in the pulse-voltage resistor-divider network. Other possible faults that could cause the same picture symptoms would be a defective Q912 blanking transistor or an open SC930 diode and R939 resistor.

Picture Symptom: Picture and sound goes off intermittently with no set pattern. Look for trouble in the power supply and particularly the limit switch shut down circuit. If the set comes back on after being turned off a few minutes, this indicates that the limit switch has shut off the set due to an increase in high voltage or a fault in the limit switch.

Circuit Operation: This regulated power supply (Fig. 3-7) does not utilize voltage feedback. Instead, a zener is used as a voltage reference. With zener CR215 in the input, Q214 merely acts as a constant voltage generator. With the emitter of Q214 at a fixed voltage, the current has to be amplified to supply enough current to Q215. Q213 is the second current amplifier. To better understand this operation we can replace zener CR215 with a 120-volt battery having a 1-ohm impedance. Thus if the current drawn from this battery does not exceed 0.5A, the voltage drop will not be greater than 0.5V.

Transistor amplifiers Q213 and Q215 each have a current gain of 30 (minimum). With both current gains multiplied, the total gain will be over 900. Now assume the maximum current through Q215 equals to 450 mA. With a gain of 900, the current drawn from Q213 in the base will be 450/900, or 0.5 mA. This

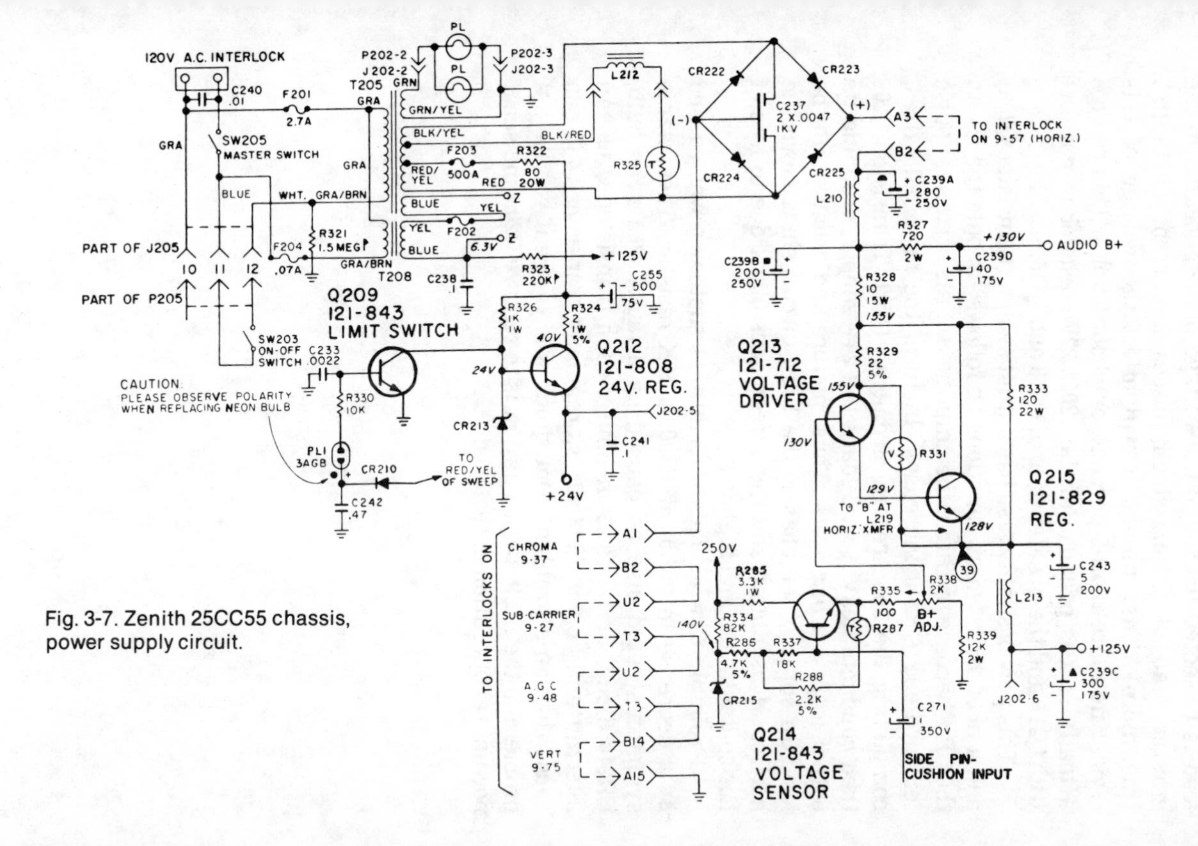

Fig. 3-7. Zenith 25CC55 chassis, power supply circuit.

load is reflected back into the voltage divider located in the emitter lead of Q214, which passes a current of 10 mA (minimum). That means it will affect the voltage of this divider no more than 5 percent, or about 6 volts. This change actually is less because of the 120Ω 22W resistor in parallel with Q215 and the greater gain of Q213 and Q215.

This power supply also has a limit switch (transistor Q209) and a neon bulb which functions as follows: Should the HV to the CRT become too high, the amplitude of the horizontal pulse fed to diode CR210 increases. In turn, the neon bulb fires, completing the base return circuit for Q209. Transistor Q209 then conducts heavily and loads the 24V supply, dropping its output to just a few volts. Thus, picture and sound would be lost. On some other chassis the limit circuit shuts down the horizontal oscillator and the screen goes black due to loss of horizontal sweep and high voltage.

Probable Cause: This intermittent shutdown was caused by a decrease in value of R330, a 10K resistor. These same symptoms could also be caused by a faulty PL1 neon bulb, CR210 diode, or leakage in limit switch transistor Q209. An intermittently open R333 across regulator transistor Q215 can cause the same problem. One quick way to see if you have a problem in the limit switch circuit is to remove Q209 and see if normal set operation is restored.

Horizontal Symptoms

This chapter includes those symptoms that give an indication—at least at the first glance—of being traceable to the horizontal circuitry. The classic symptom, of course, is streaking lines that appear as an off-frequency horizontal oscillator. If the symptom is not covered in this chapter, check some of the others. Many horizontal instability problems are accompanied by a failure of the sync circuits, which involves the vertical circuitry as well. Check the chapters on bending, vertical faults, etc., since these symptoms are quite often interrelated.

In this chapter, as in the others, set problems are covered according to brand name (manufacturer); listings are alphabetical. Bear in mind that in the vast majority of cases the brand of the set makes little difference. The problems in one receiver generally apply to all sets of similar circuitry, regardless of make or model year.

Picture Symptom: Complete loss of vertical and horizontal lock. Look for trouble in the horizontal processor and 1C count down circuits.

Circuit Operation: Shown in Fig. 4-1 is the horizontal processor circuit, IC801, which is used to develop a positive-going 31.5 kHz pulse used for horizontal deflection, after a 2-to-1 countdown in IC800 (Fig. 4-2). The 31.5 kHz pulse is also counted down in IC800 to 60 Hz for vertical deflection.

The composite sync signal is fed to terminal 3 of IC 801 via R814 and C815. A horizontal sample pulse from the sweep transformer is coupled to terminal 4 via R826 and C816. The output (positive-going 31.5 kHz pulses) is found at terminal 1.

A horizontal phase detector is used in IC801. If there is a frequency or phase shift between the sample and sync pulses, correction occurs at output terminal 1 to keep the 31.5 kHz in sync with the composite sync signal.

Horizontal frequency is adjusted by R817 (horizontal hold adjustment) which, in conjunction with R818, R821, and R822, forms a voltage divider network. The horizontal hold control is adjusted with the input sync shorted to ground at TP800. The 31.5 kHz positive-going output pulses at terminal 1 are coupled to the countdown IC where they are divided by 2 to produce the horizontal sweep signal. They are also applied to a counter string to produce the 60 Hz vertical-sweep sync signal.

This chassis has a countdown circuit that takes the horizontal signal from the IC801 processor and counts it down to the vertical scan rate. The output of IC800 provides negative-going pulses at the vertical sweep rate. This output drives predriver stage Q600 and starts the saw-forming action of C600. IC800, therefore, takes the place of a vertical oscillator circuit, and since the vertical signal is derived from the frequency-fixed horizontal signal, this set requires no vertical hold control.

The frequency-doubled horizontal signal from the output of horizontal processor IC801 is coupled to terminal 1 of IC800. A flip-flop circuit in IC800 divides this signal by 2 and couples it out of IC800 at terminal 2 to the base of horizontal driver Q800. This signal is pulsed at the horizontal sweep rate.

The incoming 31.5 kHz singal at terminal 1 is also fed to a counter string consisting of the 10 flip-flop circuits in IC800. Here the 31.5 kHz signal is counted down to the 60 Hz vertical

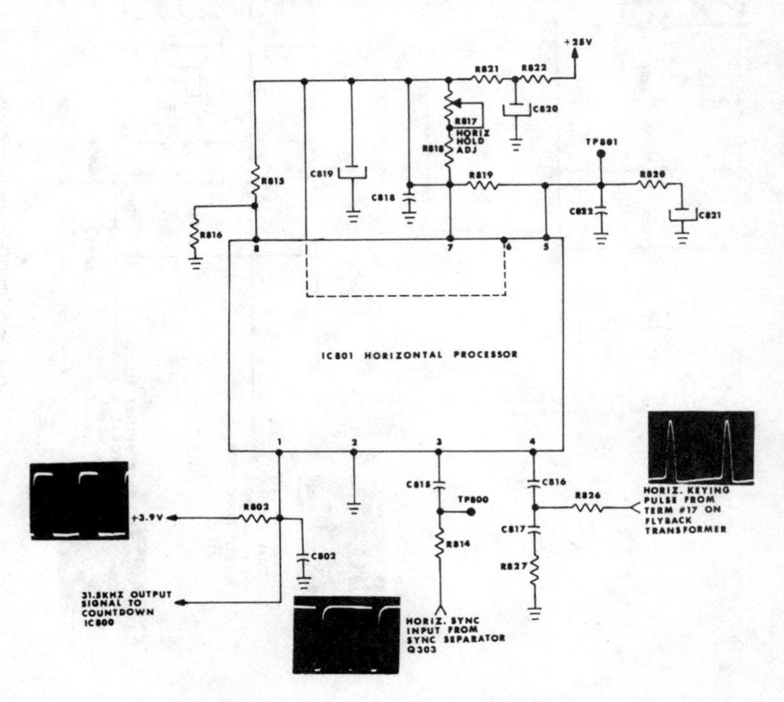

Fig. 4-1. Admiral horizontal processor circuit.

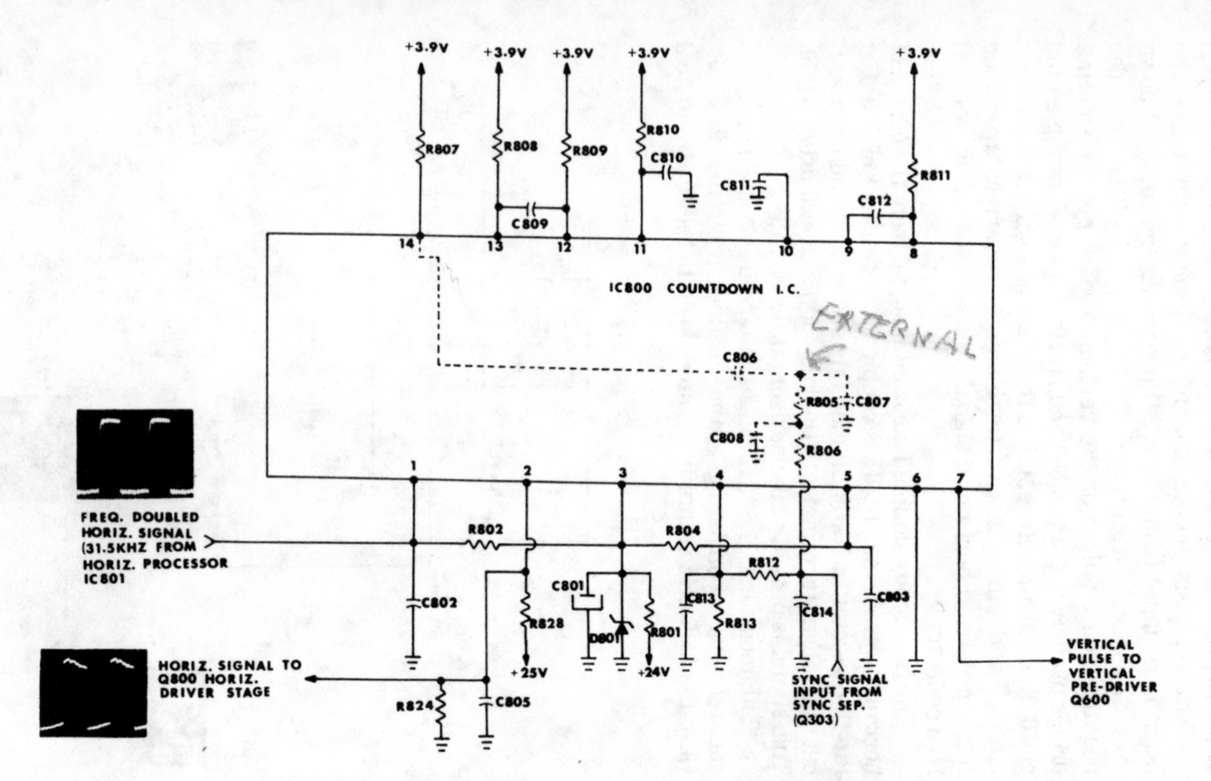

Fig. 4-2. Admiral countdown circuit for deriving vertical sweep signal.

sweep rate. This results in negative-going pulses at the vertical rate at pin 7 of IC800. This derived signal is fed to the vertical predriver stage Q600, where saw-forming and amplification takes place. The resulting pulse starts the vertical retrace.

To synchronize this operation, the negative-going vertical sync signal from the sync separator Q303 is coupled into terminals 4 and 14 of IC800. All external components relating to these terminals provide the dividing and waveshaping required to meet the input requirements of IC800.

Under normal conditions (vertical sync signal in phase), the counter initiates retrace on horizontal line number 525, and this provides a vertically locked-in picture. During poor signal conditions (as with weak or absent vertical sync equalizing pulses), the counter will initiate retrace from the vertical sync signal.

With the vertical sync signal out of phase (as during a channel change), the counter initiates retrace regardless of correct count (as during a no-signal condition).

Under a no-signal condition (no vertical sync signal at pin 7), vertical retrace (in time sequence with the leading edge of the output pulse) is started on horizontal line number 541, to automatically maintain vertical sweep.

B+ for IC800 (at terminal 3) is supplied from the +24V line via R801, and it is regulated by zener D801 to +3.9V.

Probable Cause: This complete loss of vertical and horizontal sync lock was caused by an open C815 sync coupling capacitor at pin 3 of horizontal processor IC801. Use your scope to check for a loss of sync pulses. A loss of the horizontal keying pulse at pin 4 of IC801 will cause the same picture symptoms.

Picture Symptom: No raster or high voltage, but the set has normal sound. Check the buffer module and B+ filter capacitor system.

Circuit Operation: The buffer module (Fig. 4-3) contains the horizontal buffer stage and other parts that are functional only when connected to other circuits. These components include CRT bias adjustment R1133 at pin 7, the attenuated keying pulse for the AGC at pin 13, the horizontal centering control at pin 8, and the −21V scan-rectified voltage supply for the RGB module.

Horizontal pulses from horizontal oscillator Q1007, on the signal conditioner module, drive the base of buffer Q1101, which acts as a switch when it goes from cutoff to saturation. The switching of the buffer produces a 150V peak-to-peak pulse, which is transformer coupled by T1101 to provide current drive at the base of horizontal output transistor Q1701.

Probable Cause: If a new buffer module restores the picture and high voltage, check to see if R1103 is burnt up on the old module. If this resistor is burnt, check for a loose connection (cold solder joint) on filter capacitor C1702B on the main chassis; this will be a brown lead. This filter capacitor for the +130V supply line could also be faulty (open).

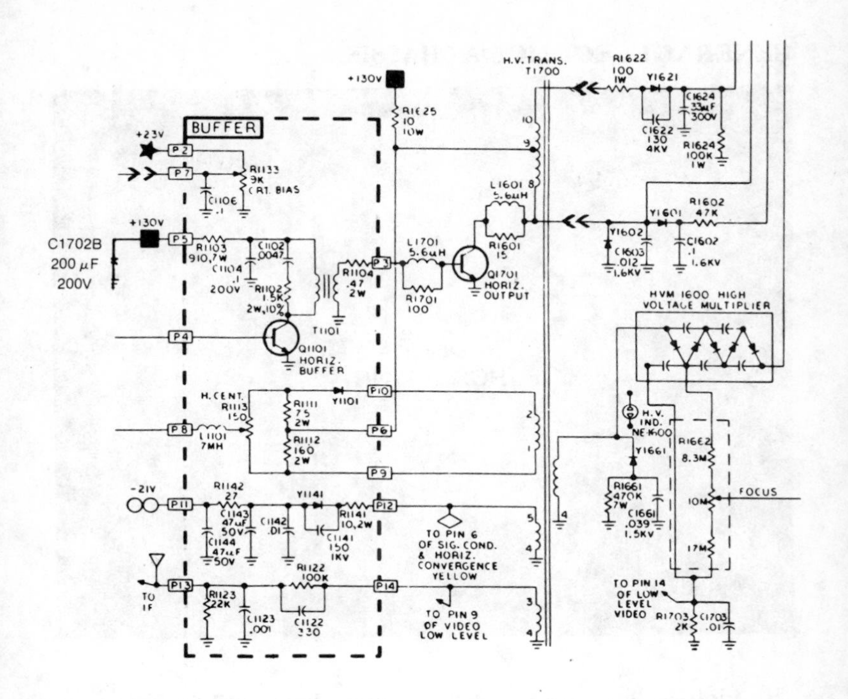

Fig. 4-3. GE MA/MB//MC chassis, horizontal buffer circuit.

Picture Symptoms: Picture shifts back and forth across the screen, and at times has a small horizontal flutter. Look for trouble in the *signal conditioner* module, around the horizontal AFC and oscillator control circuits.

Circuit Operation: For horizontal sync, the output of sync separator Q1003 is fed to the phase detector of the AFC system, which compares the frequency of the horizontal oscillator with the frequency of the sync, using the horizontal reference pulses. Refer to module circuit in Fig. 4-4. When these two signals differ in frequency and phase, the AFC produces a correcting DC voltage to bring the horizontal oscillator back on frequency and with the proper phase.

The output of the phase detector is fed to reactance transistor Q1006, through a control network that consists of RC filter R1036, C1018, and C1019. This circuit controls horizontal pull-in, hold-in, and stability during weak signals and strong noise conditions. The RC control network provides a signal to reactance transistor Q1006, which varies the frequency of Hartley oscillator Q1007, according to the AFC correction voltage. The oscillator output is then shaped and fed from pin 8 to horizontal buffer Q1101.

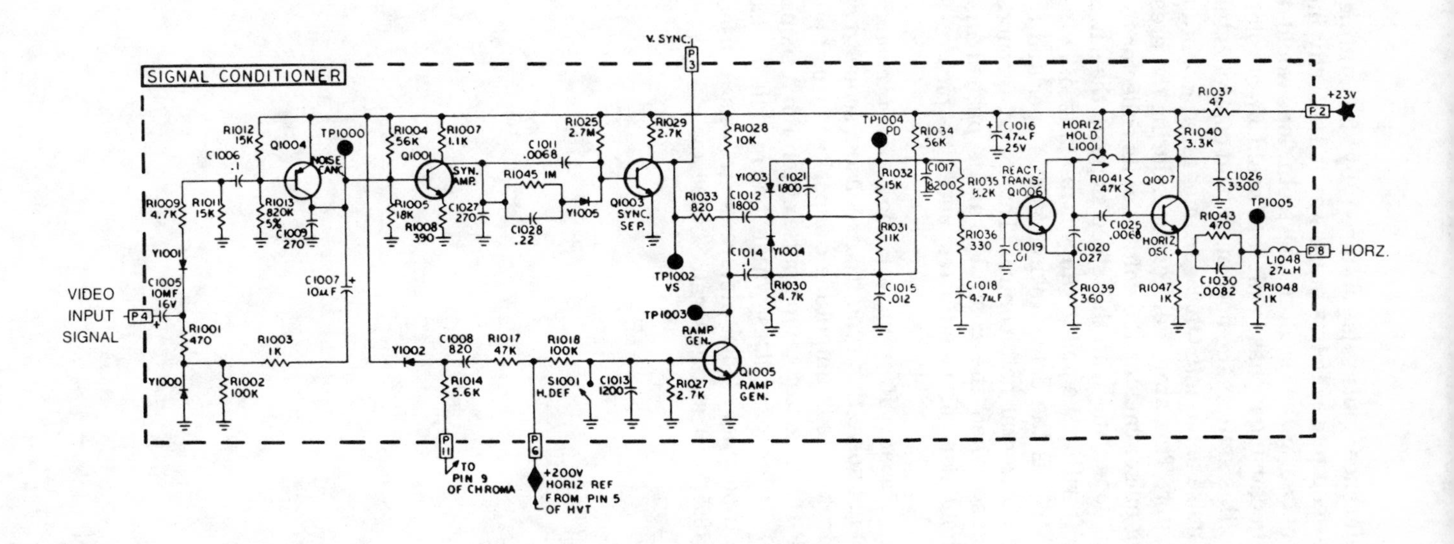

Fig. 4-4. Signal conditioner module in GE MA chassis.

The reference pulse is obtained by integrating the horizontal sweep pulse. If this pulse was fed directly into an integrating network, the resultant sawtooth slope would not be steep enough, causing a phase shift that would affect color hue. To minimize this shift as the horizontal hold control is turned, the flyback pulse is fed into Q1005, an integrator amplifier and ramp generator. This amplifier conducts during the presence of the keying pulse, causing the voltage at its collector to drop with a steep slope. And with this steep slope on the sawtooth, the phase shift within the pull-in range is minimized.

Switch S1001 is the defeat switch and is operated by the shaft on the horizontal hold control. This switch grounds the reference pulse when the control shaft is pushed in. The horizontal hold control is adjusted until the picture stops floating.

Probable Cause: If a known good *signal conditioner* module does not solve this sync trouble, use a scope to check for correct peak-to-peak amplitude composite video at pin P4, and a 200V P-P horizontal keying pulse at pin P6. A scope can also be used to check pulses within the module to isolate defective components. This picture shift was caused by leakage in Q1005, the ramp generator transistor.

Picture Symptom: No picture and no high voltage. Sound okay. Circuit breaker is open, and when the breaker is reset, there is no sound or picture until the breaker trips after about 15 seconds. Then the sound comes on again. Look for a short in the +142 volt line.

Circuit Operation: A short in the +142V line could be caused by horizontal output transistor Q1701, damper diode Y1602, or saturable reactor transformer, T1704. Because this trouble was in the high-voltage regulation system, let's look at that circuit operation.

High-voltage regulation is controlled by T1704, a saturable reactor that has three windings (Fig. 4-5). The control winding is in series with the primary of the T1700 and the +143 volts DC supply to the collector of Q1701, the horizontal output transistor. The shunt winding is in parallel with the winding, and the series winding is in series with the horizontal deflection yoke.

When current flow is increased through the control winding, saturation increases in the core and the inductance of each of the three windings is reduced. If current is reduced, the inductances will increase.

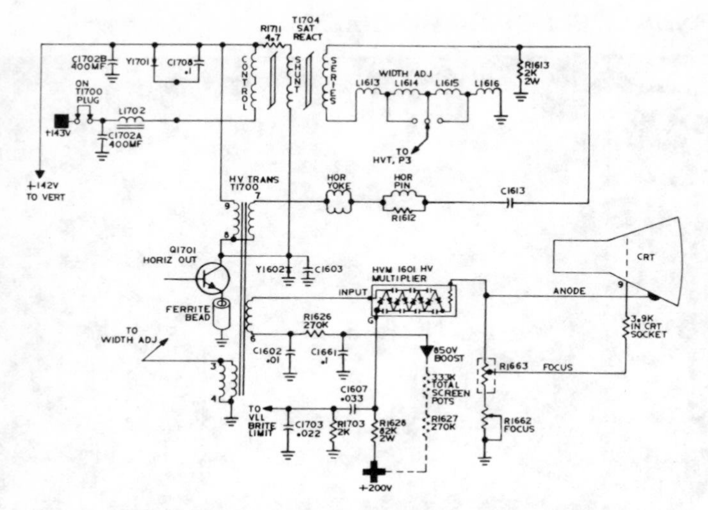

Fig. 4-5. GE MC chassis, sweep output system.

When the load on the system is increased by increasing brightness, current flow is increased through the primary and the control winding of T1704. The increase in current increases saturation in the core, thus lowering the inductance of the shunt winding. This lowers the total primary circuit inductance since the primary and shunt windings are in parallel, and the lower primary circuit inductance shortens the retrace time. This narrows the flyback pulse and increases its amplitude to maintain a constant high voltage.

The inductance of the series winding of T1704 is also lowered when the core saturation is increased, and the lower impedance in the yoke circuit increases the yoke current to maintain correct sweep width. When brightness is reduced, the load on the horizontal system is decreased and the opposite action occurs—saturation of the core is less, inductances increase, retrace time increases to reduce the flyback pulse amplitude, and a constant high voltage is maintained.

Due to capacitive effects in the saturable reactor, the control winding has a tendency to ring. C1708 lowers the resonant frequency of the control winding to minimize ringing and Y1701 clamps any residual pulses.

The +143V supply is filtered by C1702A, C1702B, and choke L1702. An interlock, between the bridge rectifier and the filter, is opened whenever the HVT socket is unplugged. This feature protects Q1701 from having 143V applied to its collector through the shunt winding of T1704. R1711 limits current through the shunt winding. The ferrite bead on the emitter lead of Q1701 suppresses any ringing that might occur on the left side of the screen.

The boost voltage is regulated by selecting the correct value of R1628 (82K, 2W). Beam electron flow is through two paths. One path is through the anode, the HV multiplier, the HV winding, R1626, all of the screen pots, and R1627. The other path is through the anode, the HV multiplier, the clamp diode, and R1628. Since the resistance of R1628 (82K) is much smaller than the total resistance (873K) of the other path, most of the beam current flows through R1628. Therefore, when the beam current increases the boost voltage increases. But the B+ voltage added to this boost voltage is reduced by the voltage drop across R1628 caused by the increased beam current so the total boost voltage is regulated.

Probable Cause: This loss of high voltage was due to a shorted (grounded) coil of saturable reactor transformer T1704. Other possible faults that cause loss of HV and a short on the +142V line include components Q1701, Y1602, and C1702A, and C1702B filter capacitors.

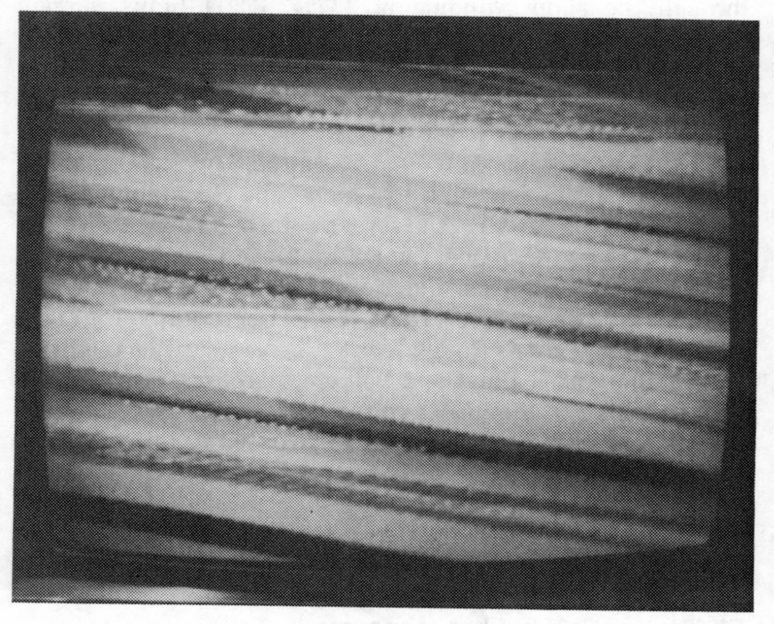

Picture Symptom: Diagonal lines across the screen, which indicates incorrect horizontal oscillator frequency or loss of horizontal sync. Look for trouble in the horizontal AFC or oscillator circuits.

Circuit Operation: The horizontal AFC circuit (Fig. 4-6) compares the sync pulses from the plate of the sync separator with a sawtooth waveform from the horizontal oscillator. If the two are of different phase, one diode conducts more heavily than the other and feeds a correction voltage to the DC amplifier, which changes the bias voltage on the grid of the oscillator and corrects the phase relationship.

A dual-triode 6FQ7 (V502) is used as a combination DC amplifier and horizontal blocking oscillator. As the DC amplifier conducts, a bias voltage is developed across horizontal hold control R108 and R107. This voltage, applied to the grid of the oscillator section, starts the oscillation. L501B and C528 consititute the ringing circuit. C529 acts as the shaping capacitor, and a portion of the waveform across this capacitor is coupled back to the horizontal AFC circuit through C524 for oscillator frequency comparison with the horizontal sync pulses.

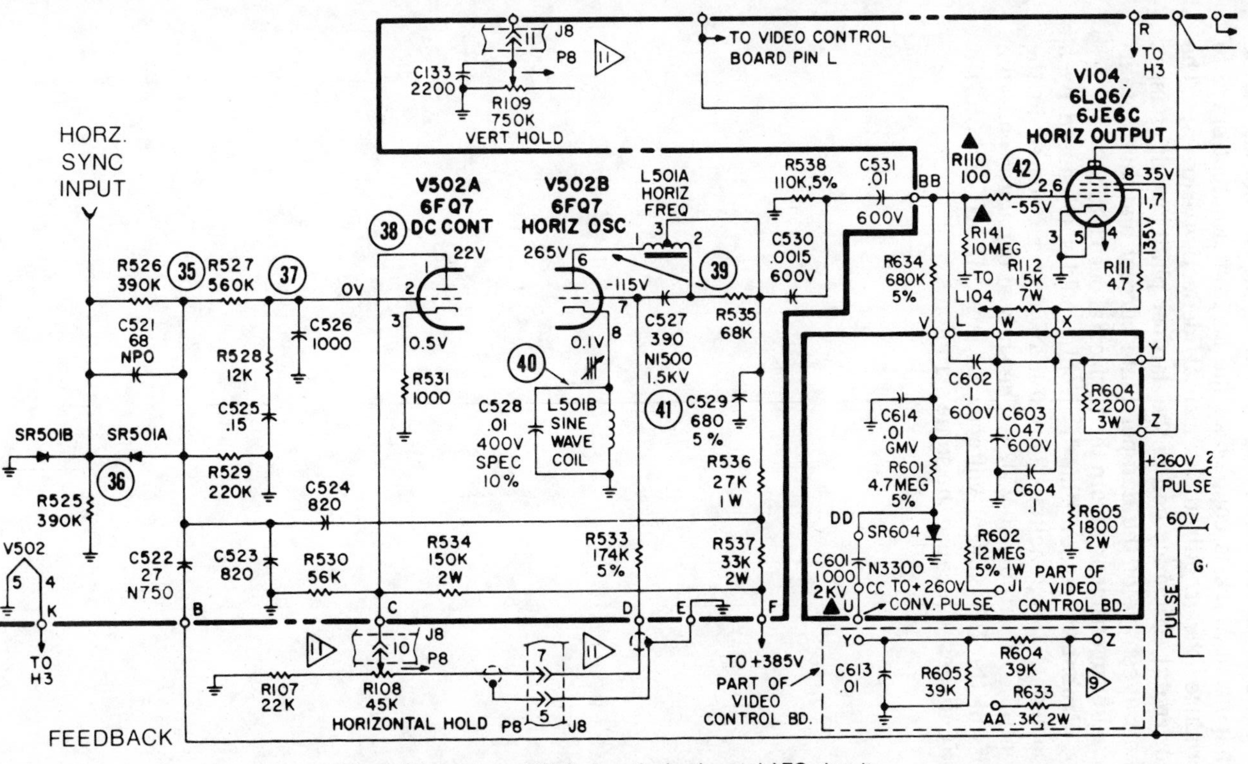

Fig. 4-6. Magnavox T940 chassis, horizontal AFC circuit.

The modified sawtooth appearing at the output of the shaping network is coupled to the horizontal output stage to produce the required horizontal deflection current. The horizontal efficiency coil in the damper plate circuit changes the waveshape of the damper plate current and thus obtains maximum efficiency from the horizontal output tube.

Probable Cause: This horizontal hold was caused by a shorted C525 capacitor. This 0.15μF unit is located in the control grid circuit of V502A. The shorted capacitor caused a loss of the DC control voltage. A faulty SR501B and SR501A AFC diodes will cause the same symptoms.

MOTOROLA (QUASAR) TX-942 CHASSIS

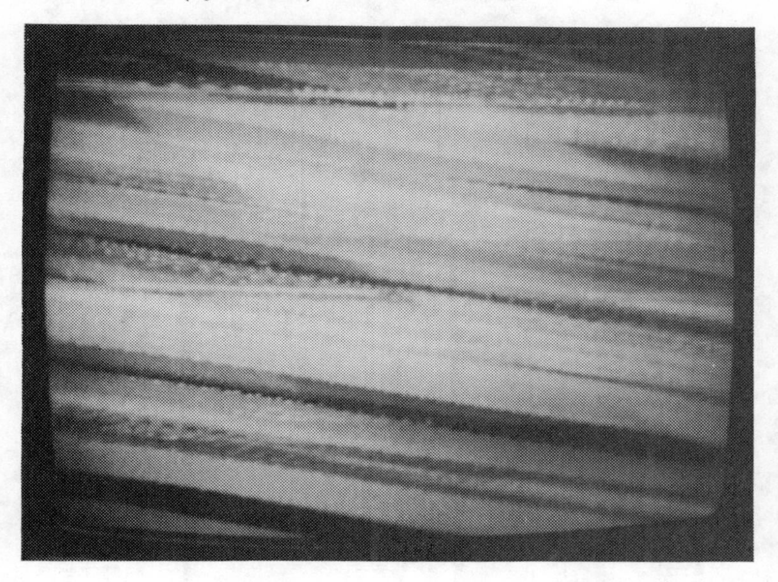

Picture Symptom: Loss of horizontal sync. Look for trouble in the horizontal FC panel or associated circuits that supply sync and keying pulse signals to this panel.

Circuit Operation: The FC panel (Fig. 4-7) contains the horizontal AFC, sweep oscillator, pulse former, and a mini-mode power supply. Horizontal sync enters the panel at pin 14 and is applied to both the color-gate pulse former and the horizontal phase detector.

Color-gate pulse former Q1 has a negative-going horizontal sync pulse at its base and a horizontal output pulse on its collector during retrace time. This transistor is normally saturated and is cut off momentarily by the sync pulse on its base. The resulting output at this time is a positive pulse of the proper timing and duration to gate the color sync stage on the chroma IC, which is located on the SC panel. Thus, a faulty color gate transistor (Q1) will cause a loss of color.

Horizontal output and horizontal sync pulses are compared in the phase detector circuit. Any difference in time (phase) between the pulses results in the development of a correction voltage that keeps the horizontal oscillator in proper synchronization. Horizontal oscillator output is fed to a buffer stage for isolation. The waveform is amplified by the

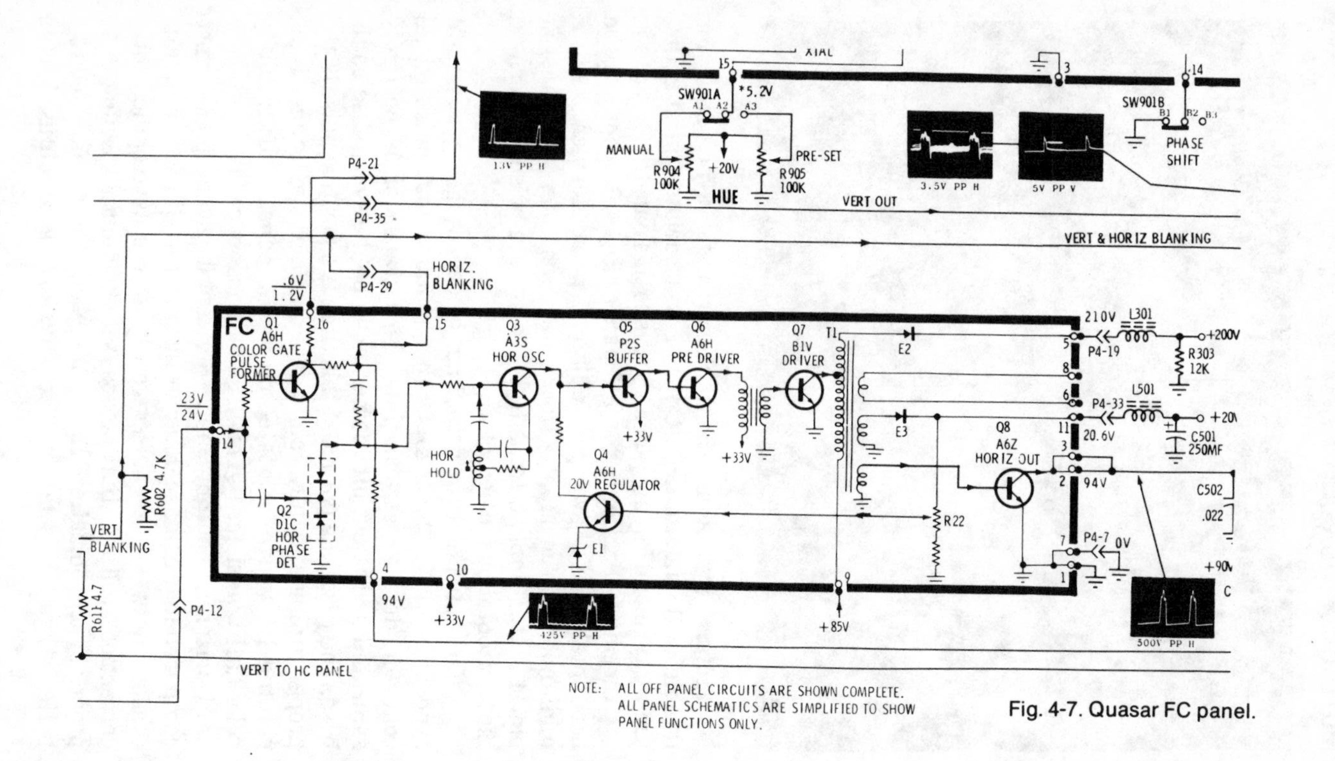

Fig. 4-7. Quasar FC panel.

pre-driver and driver stages and then transformer-coupled to the output stage. The driver transformer also provides two other voltages which, after rectification and filtering, provides 20 and 200 volts for other portions of the receiver. The 20V supply is closely regulated by varying the pulse width of the oscillator.

Probable Cause: Test by plugging in a known good FC horizontal module. Possible suspect components that could cause loss of horizontal sync on the module would be pulse-former transistor Q1 and horizontal phase detector Q2. Use an oscilloscope to check for the horizontal keying pulse at pin 4 and sync pulse at pin 14.

This sync problem was caused by a faulty (open) connection at pin 14 of the FC module. Also, do not overlook the sync circuits found in the VA panel.

Picture Symptom: Dark screen and no high voltage at CRT anode. Look for trouble in the horizontal deflection system.

Circuit Operation: The RCA CTC-40 chassis generates the desired horizontal sweep current with circuitry using silicon controlled rectifier (SCR) devices and associated circuits. Note the simplified circuit in Fig. 4-8.

Essentially, diode D1 and controlled rectifier SCR1 provide the switching action that controls the current in horizontal yoke windings LY during the picture-tube beam-trace interval. Diode D2 and controlled rectifier SCR2 control yoke current during the retrace interval. Components LR, CR, CH, and CY provide the necessary energy storage and timing. Inductance LG1 supplies a charge path for CR and CH from the B+, thereby providing a means to "recharge" the system from the set's power supply. Inductance LG2 provides a gating current for rectifier SCR1 (LG1 and LG2 comprise transformer T-102). Capacitor CH optimizes the retrace time by virtue of its resonant action with LR.

Sweep Circuit Quick-checks: Loss of horizontal sweep and high voltage may or may not cause a power supply overload, depending on the specific trouble. A faulty high-voltage

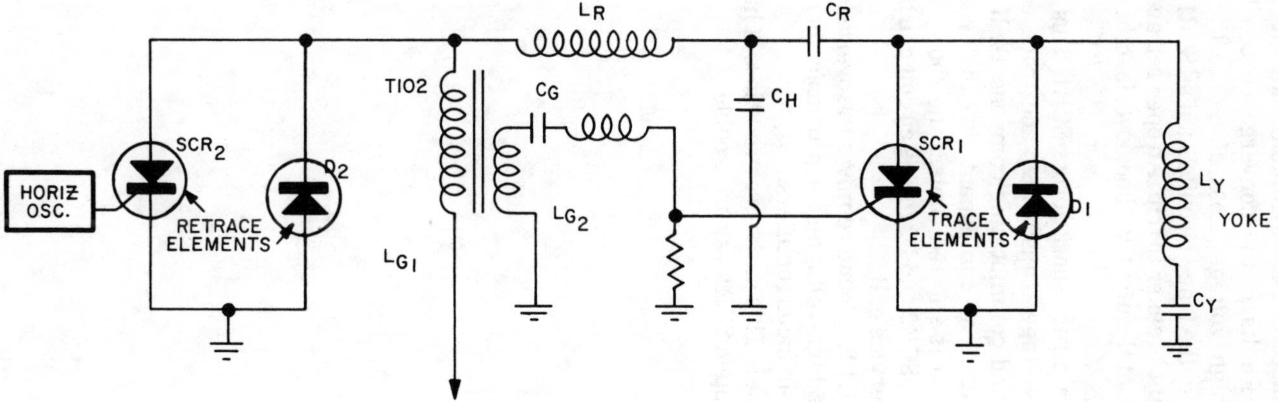

Fig. 4-8. RCA CTC-40 SCR horizontal sweep circuit.

rectifier can cause either symptom and it should be substituted first. Faults in the convergence board will also cause a loss of high voltage. Loss of B+ to the deflection circuit, or a short in the trace circuits such as SCR1 or D1, will not cause the circuit breaker to trip. An open retrace SCR2 will also kill all horizontal output without overloading the power supply.

Now connect a direct short across SCR1. This effectively isolates the high-voltage transformer and trace components from the retrace and commutating components. If the circuit breaker still trips, a component in the retrace and commutating circuit is shorted (unless the short is not in the deflection system). SCR2 and D2 in this circuit can be replaced during the home service call.

If shorting SCR1 to ground removes the overload, the fault probably is in the high-voltage sweep transformer circuitry. Check the two slip-in focus rectifiers.

Probable Cause: This dark screen and loss of high voltage was caused by an open SCR1—a rare trouble.

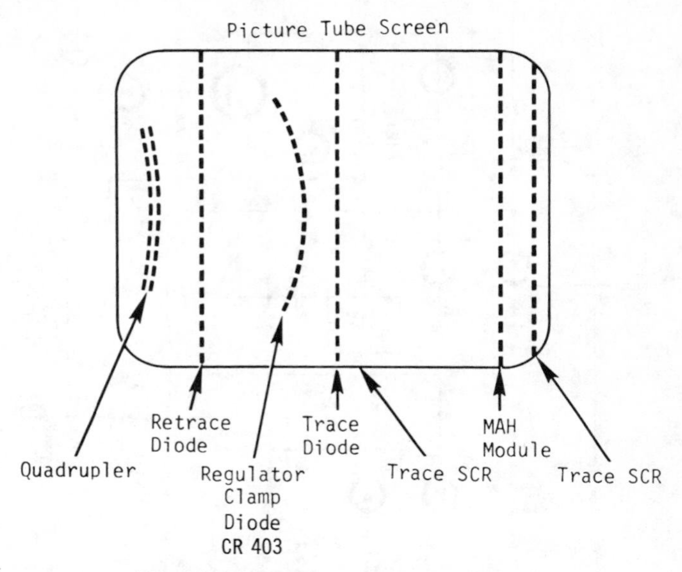

Fig. 4-9. Interference patterns on screen.

Picture Symptom: Interference lines or patterns on the screen (see Fig. 4-9) during reception of weak TV station signals. Look for switching transients in the horizontal sweep system.

Circuit Operation: The *trace* resonant circuit (Fig. 4-10) consists primarily of C408 and C409 in parallel with T404, the yoke, L404, and either CR-101 or SCR-101, whichever is conducting. The yoke has the largest inductance. The resonant frequency of this circuit is about 10 kHz, and the half-cycle period is nearly equal to one duration of the scanning interval.

The *retrace* circuit is made up of C408 and C409 in parallel, T404, the yoke, L404, C413, and C414 in parallel with L104 and either CR-102 or SCR-102, whichever is conducting. The resonant frequency of this circuit is such that its half-cycle period is equal to retrace time.

The power-input resonant circuit consists of R424 and T402 in parallel with T403, L104, and two parallel paths to ground. One of these parallel paths is C412 and R407 paralleled by L406. The other is C413 and C414 in series with either SCR-101 or CR-101, whichever is conducting. The resonant frequency of this circuit allows the voltage at the junction of T403 and L104

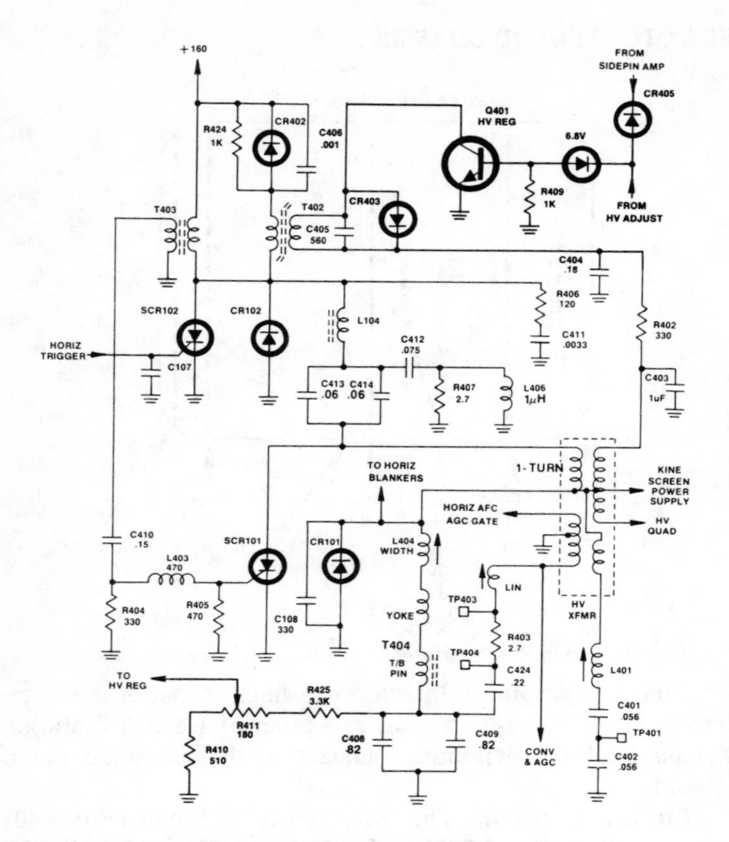

Fig. 4-10. RCA CTC-49, horizontal sweep circuit.

to rise to its positive peak and begin to decay during each scanning interval. The precise frequency is varied by the voltage regulator. Current flowing through T403 generates a positive gate voltage for SCR-101 during most of the scanning interval.

Referring to the timing diagram in Fig. 4-11, note that at time zero a large current is flowing in the trace circuit, and conduction is through CR-101. This current decays to zero at midscan and reverses, flowing through SCR-101 and rising from zero to maximum at the right end of the scan. Near the end of the scanning interval, gate voltage is removed from SCR-101, but it continues conducting.

Throughout the scan interval, the lower terminals of C413 and C414 are grounded. Electron current flows from ground through C413, C414, C412, and the rest of the power-input circuit to B+. L104 is small compared to T403 and T402, so the voltages at its opposite terminals are about the same. The voltages at the anode of SCR-102 and CR-102 reaches maximum slightly before the end of the trace and begins to decay, since the power-input circuit is resonant. Regulation of input power is accomplished by changing the resonant frequency, which is a function of saturable reactor T402.

When SCR102 is triggered, the retrace resonant circuit is completed, and the energy stored in C412, C413, and C414 causes electron flow to continue downward through the yoke, rapidly cutting off SCR-101.

Probable Cause: As we have just seen, a lot of switching transients could and do occur in this sweep system, and these switching transients may show up as interference patterns on weak stations. The general location and shape of the interference pattern on the screen can be a tip-off as to which component should be replaced to eliminate the interference.

Refer back to the CRT screen patterns in Fig. 4-9. The interference caused by regulator clamp diode CR-403 may appear as a straight line (rather than bowed) in some chassis. Part No. 131475 (trace diode) or No. 131476 (retrace diode) can be used as a replacement for CR-403 in this chassis.

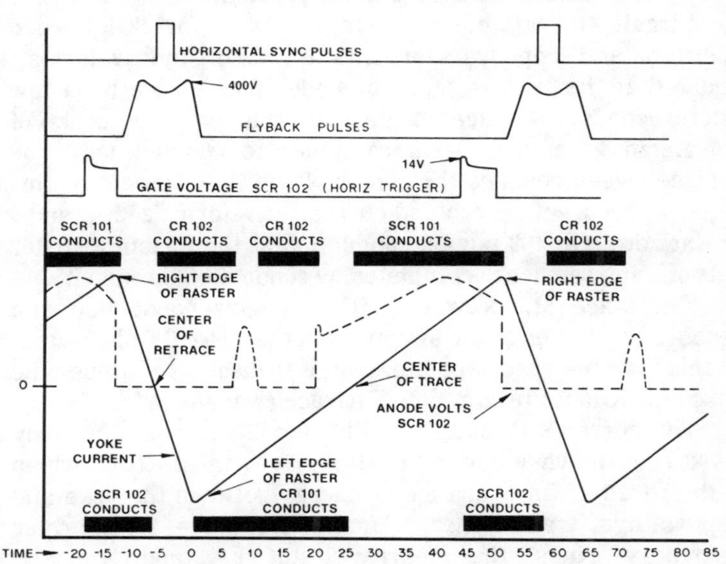

Fig. 4-11. Timing diagram of horizontal deflection.

Picture Symptom: When the picture brightness level was very high, the circuit breaker would trip out. Look for trouble in the SCR horizontal sweep system. Replacement of the SCRs and damper diodes did not solve this problem.

Circuit Operation: Refer to Fig. 4-12 for SCR sweep operation and scope waveforms. You will note in waveforms 1 and 2 that the trigger pulse precedes the retrace by a few microseconds. Retrace begins at the positive peak of waveform 2, and SCR-102 continues to conduct until the retrace sweep reaches the center of the CRT screen. At this time, yoke curent is zero and current waveforms 2, 3, 4, and 5 are at zero. CR-402 begins conduction at the instant SCR-102 cuts off, and retrace is completed by conduction through it.

The trace starts when CR-401 begins conduction and reaches the center of the sweep when the current falls to zero. At this time the yoke capacitor begins to discharge through the yoke and SCR-101 to complete the trace (waveform 5).

The currents through SCR-101 and CR-401 contain many peaks, but the yoke current itself is a smooth sawtooth, which is the result of an exchange of energy between the yoke and high-voltage transformer. Current from the transformer secondary "rings" the linearity circuit (L402 and C413) and flows through the yoke, alternately aiding and bucking the

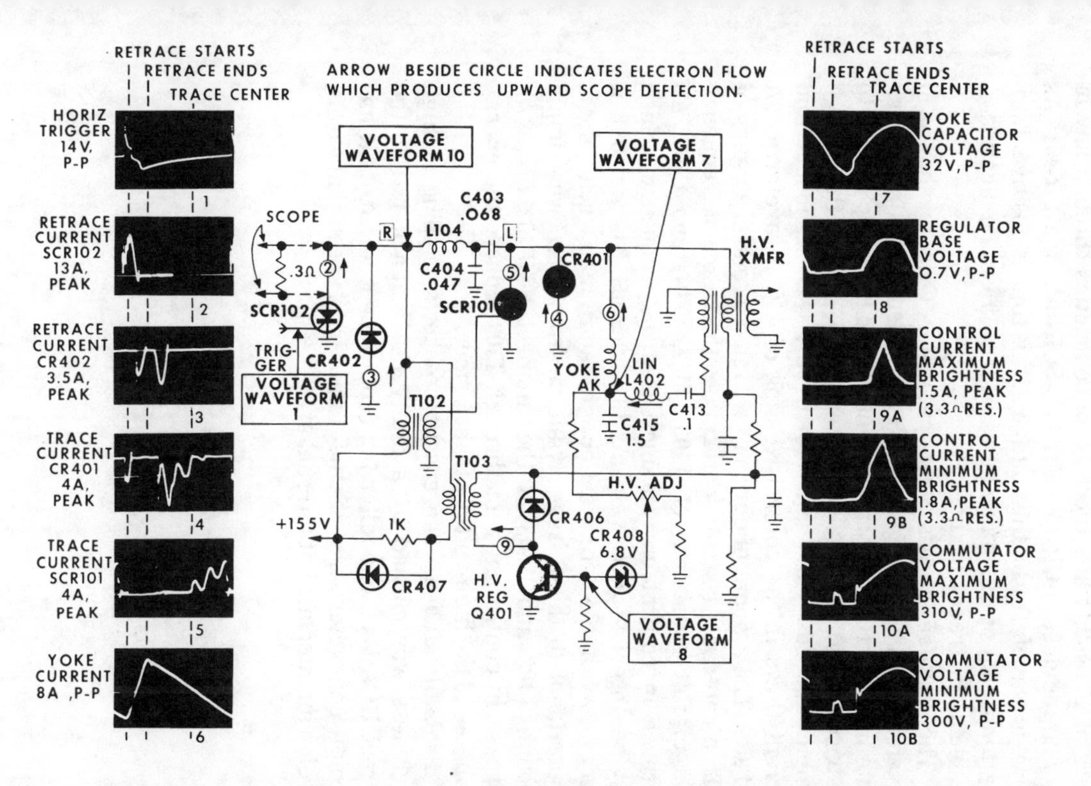

Fig. 4-12. RCA CTC-40, SCR horizontal sweep circuit.

current of SCR-101 and CR-401 as necessary to produce a linear sweep.

There is a short interval, between the two conduction times of CR-402, when there is no conduction through any of the four devices, SCR-102, CR-402, SCR-101, and CR-401. As shown in waveform 10, the voltage at this time is positive. Loss or attenuation of this pulse in waveform 10 is an indication of trouble in the trace circuit.

The voltage across yoke capacitor C415 is shown in waveform 7. The voltage at C415 is divided across a network containing the HV adjustment control, and the sample is fed to the base of Q401 via a 6.8-volt zener (CR408). When the voltage sample exceeds 6.8V (waveforms 8), Q401 conducts and current flows in the control winding of T103 (waveforms 9A and 9B). The amount of current through the control winding of T103, a saturable reactor, determines the resonant frequency of a circuit consisting of T103, T102, C404, and C403. An increase in control current increases the frequency, which may be observed in waveforms 10A and 10B. The amount of energy available for deflection and high voltage is proportional to the voltage available from the power supply through T102 and T103 at the moment SCR-102 fires. An increase in control current through T103 reduces the energy available to the high-voltage transformer.

Probable Cause: The circuit breaker's tripping at high brightness was caused by a faulty C403 capacitor. Use the correct RCA type capacitor for replacement as this unit has special characteristics. Some units have been known to develop an intermittent internal arc.

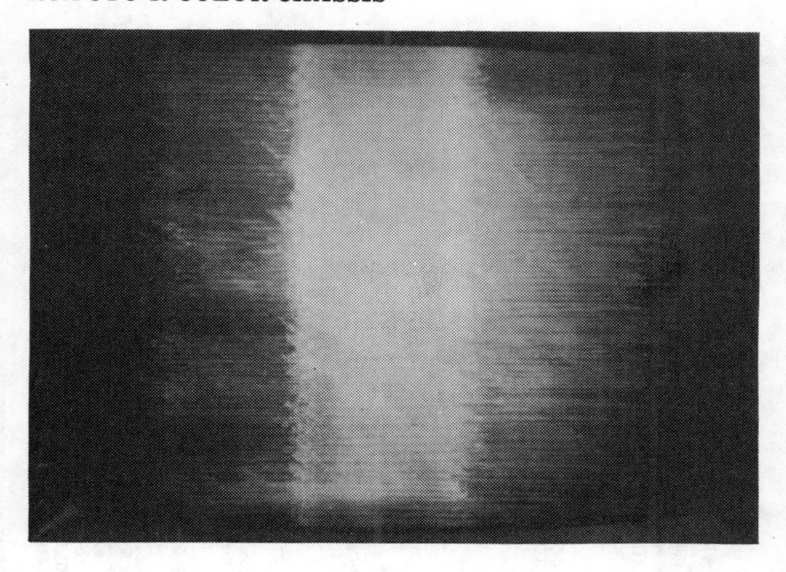

Picture Symptom: The screen of this set produced a "Christmas tree" effect, and then the circuit breaker would trip. Look for trouble in the horizontal sweep system.

Circuit Operation: Because this horizontal sweep circuit (Fig. 4-13) utilizes silicon controlled rectifiers (SCRs), it should now be helpful to review the basic operation of the SCR.

An SCR is a controlled rectifier made from the semiconductor material, silicon. The SCR action is somewhat like that of a thyraton tube—it is basically nonconductive until turned on by some control signal fed to a control electrode. When turned on, the device acts like a normal rectifier and is capable of conducting high currents and exhibiting very low forward voltage drop between its anode and cathode (Fig. 4-14). The SCR can be turned off by reversing the anode–cathode voltage or by reducing the current through it to the point where the SCR has a high resistance.

The SCR has three elements which are anode, cathode, and the control gate, as shown in Fig. 4-14. When a sufficient amount of current forward-biases the gate, the SCR will switch on. The amount of forward anode–cathode voltage that will switch the SCR into conduction depends on the level of gate current, as seen in Fig. 4-15. After the SCR has been switched

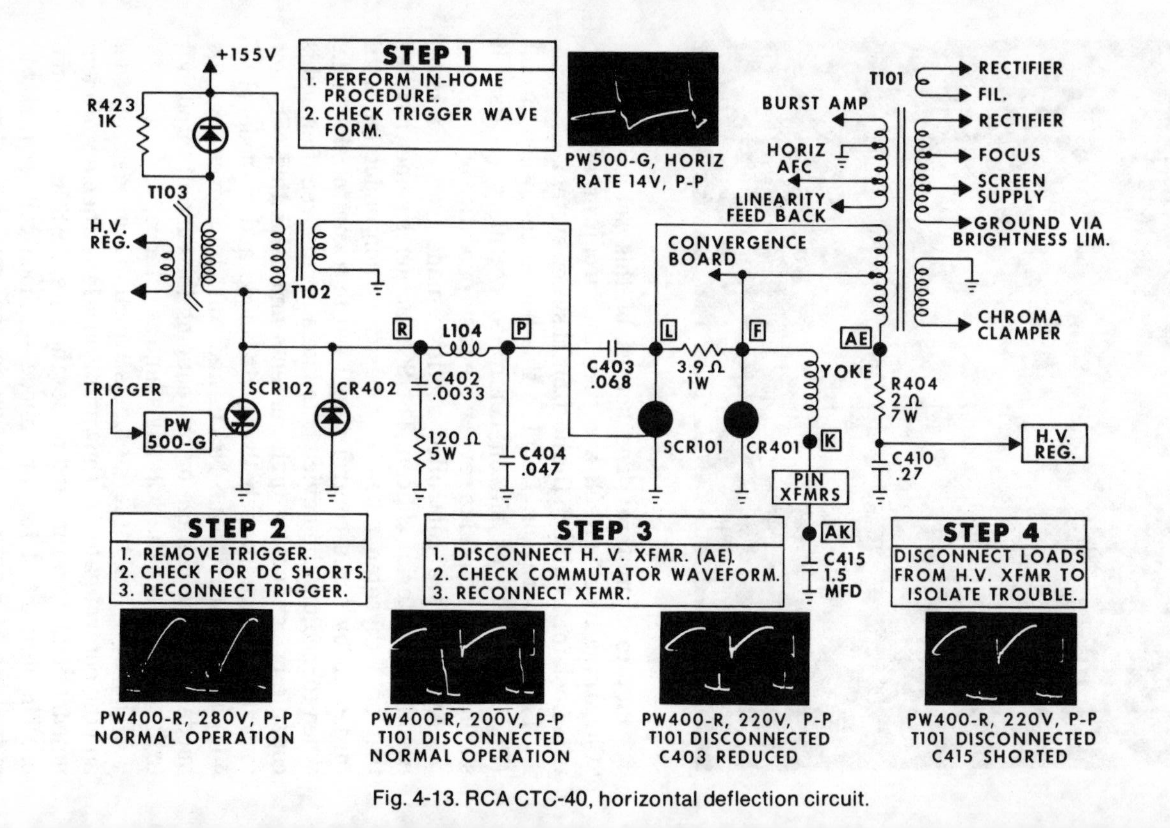

Fig. 4-13. RCA CTC-40, horizontal deflection circuit.

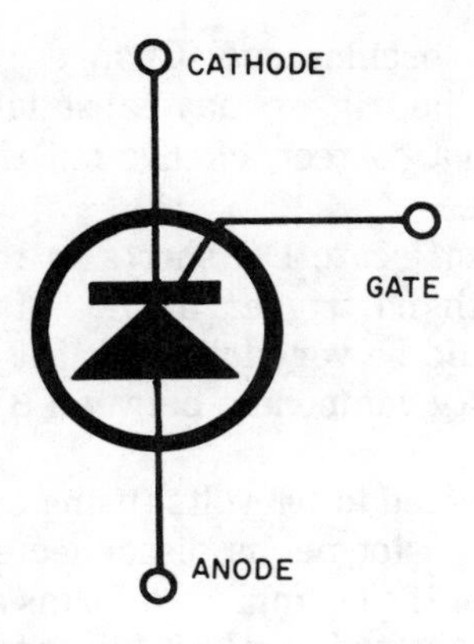

Fig. 4-14. SCR diagram.

on, the current through it is *independent* of gate current. Thus, the SCR is used to switch the horizontal sweep current.

The TV set's horizontal, electromagnetic deflection should cause a linear current to flow in the yoke windings in such a manner as to deflect the CRT beam from one side of the screen to the other. Also, the yoke current must cause the CRT beam to return to its starting point, and this is referred to as the *retrace* or *flyback* current component.

Circuit Testing Techniques: Refer to check chart and scope traces in Fig. 4-13. In step 1, check the waveshape and

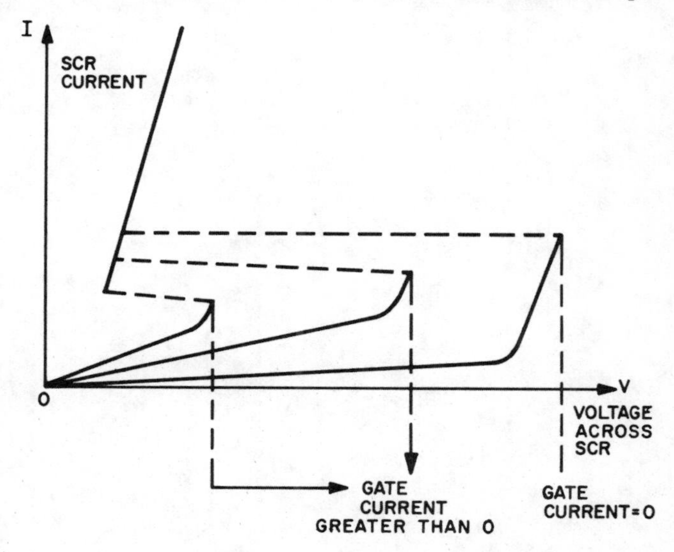

Fig. 4-15. SCR gate current relation chart.

amplitude of the trigger with an oscilloscope. Changes in amplitude, frequency, or shape of the trigger may cause loss of sweep. If this trigger pulse is not correct, change out the MAH-1 horizontal oscillator module.

Use the procedure in step 2 to locate DC shorts in the retrace commutating circuit. With no trigger applied, the deflection circuit load current should be very low, and if it is not, this indicates a shorted or leaky component between B+ and ground.

In step 3, the line voltage is reduced to 100 volts (using the "Isotap"), and the high-voltage transformer is disconnected from the sweep circuit. The shape of the normal waveforms at PW400R is important. Leaky, shorted, or off-tolerance components in the trace circuit will cause the narrow pulse contained in the waveform to be attenuated. These components include C403, C404, C415, C406, and others. Note that C410 is not checked in this step, since it is disconnected.

If trouble is not found in these three steps, the fault is probably in the high-voltage transformer itself, or in the loads connected to it. Disconnect each load until the trouble is found.

Probable Cause: This "Christmas tree" picture symptom was caused by a defective MAH-1 horizontal oscillator module.

Picture Symptom: The set has very critical horizontal lock, and the color may fade off and on or be lost completely. Look for trouble in the horizontal sync and AFC circuits located in the 9-90 horizontal module.

Circuit Operation: Negative-going sync (Fig. 4-16) is coupled to the cathodes of AFC diodes CR801 and CR802 through R801 and C801. At the same time, a negative pulse from the sweep transformer is fed to the emitter of Q804, via R808, to the sawtooth shaper transistor. The resultant sawtooth waveform is coupled to the AFC diode circuit through C802.

When the TV sync pulse and the horizontal oscillator are exactly in phase, the AFC comparator produces zero-error voltage. Should the oscillator either lead or lag the sync pulses, an "error" voltage will be produced at the anode of CR801. This error voltage can be positive or negative and will either add to or subtract from the existing positive voltage appearing at the base of horizontal AFC transistor Q801. Capacitor C807 and RC network C806-R806 comprise the AFC anti-hunt network. Q801 thus acts as a variable resistance in series with C809.

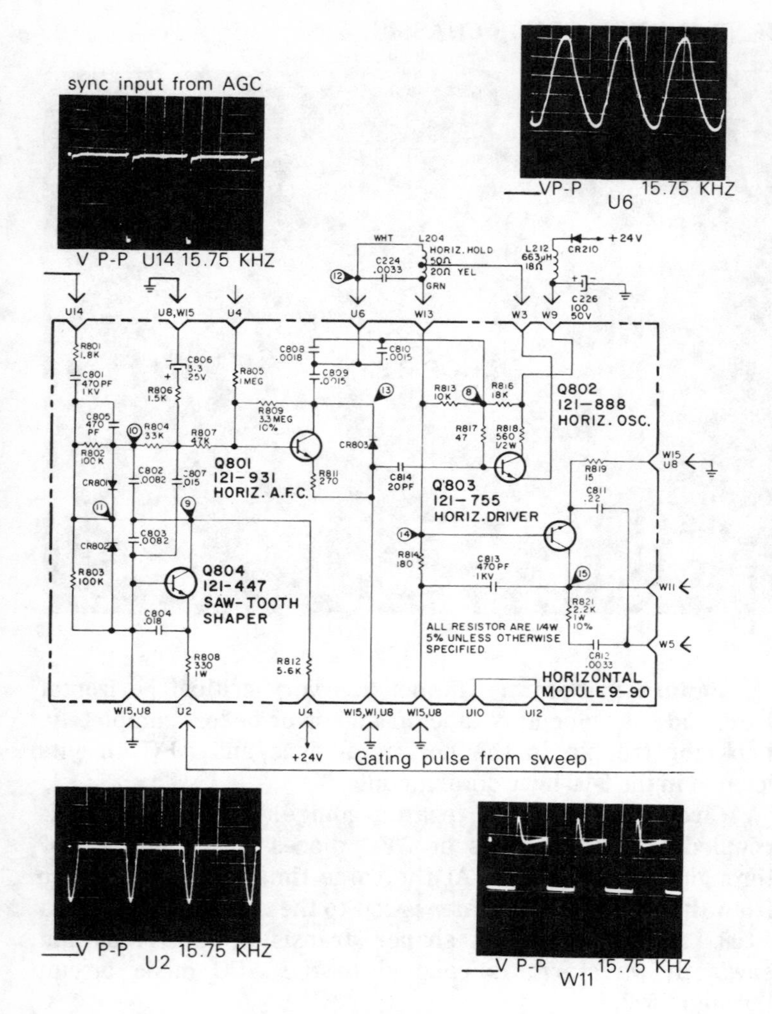

Fig. 4-16. Zenith 9-90, horizontal module circuit.

Horizontal oscillator transistor Q802 is used in a modified Hartley configuration. The resonant circuit consists of coil L204 and capacitor C224. The frequency of the circuit is determined by the setting of L204, the horizontal hold control.

The 15.75 kHz waveform generated by the oscillator is coupled to the base of horizontal driver transistor Q803. This stage in turn drives the horizontal output transistor via coupling transformer T204.

Probable Cause: Resistor R808, which feeds the horizontal comparison pulses into the emitter of Q804, was charred and had a very high resistance value. This resistor will often burn up due to the high horizontal keying pulses. The burst-keying pulses for the chroma and the keying pulses used for sync comparison in the horizontal come from the opposite polarities of the same winding of the sweep transformer.

Many times a resistor that is overloaded will go down in value, but will later on develop a very high resistance. Thus, it is always a good deed to replace any resistors that show signs of overheating. This helps eliminate any delayed resistor failures.

Picture Symptom: Dark screen. No high voltage found at the CRT anode, and no horizontal drive or sweep action found at the horizontal output transistor. Look for trouble in the horizontal driver or horizontal output system.

Circuit Operation: Now for a quick glance at the horizontal sweep circuit operation found in the Zenith "F" line chassis (Fig. 4-17).

The positive base voltage on driver transistor Q803 is derived from the horizontal oscillator stage. The horizontal pulse is fed to the base of the driver transistor and a square wave is generated in the collector circuit. The square wave (note 170V P-P pulse in top trace of (Fig. 4-18) is coupled to the base of horizontal output transistor Q202 through transformer T204.

The base of the horizontal output transistor is connected to the top of the secondary winding of horizontal driver transformer T206. The bottom of this winding goes to ground via resistor R224. Bypassing is provided by capacitor C227. The emitter of the horizontal output transistor is returned directly to ground. Ferrite bead L206 on the collector lead suppresses "spooks" generated by rapid switching of the transistor junctions. The collector is connected to the primary

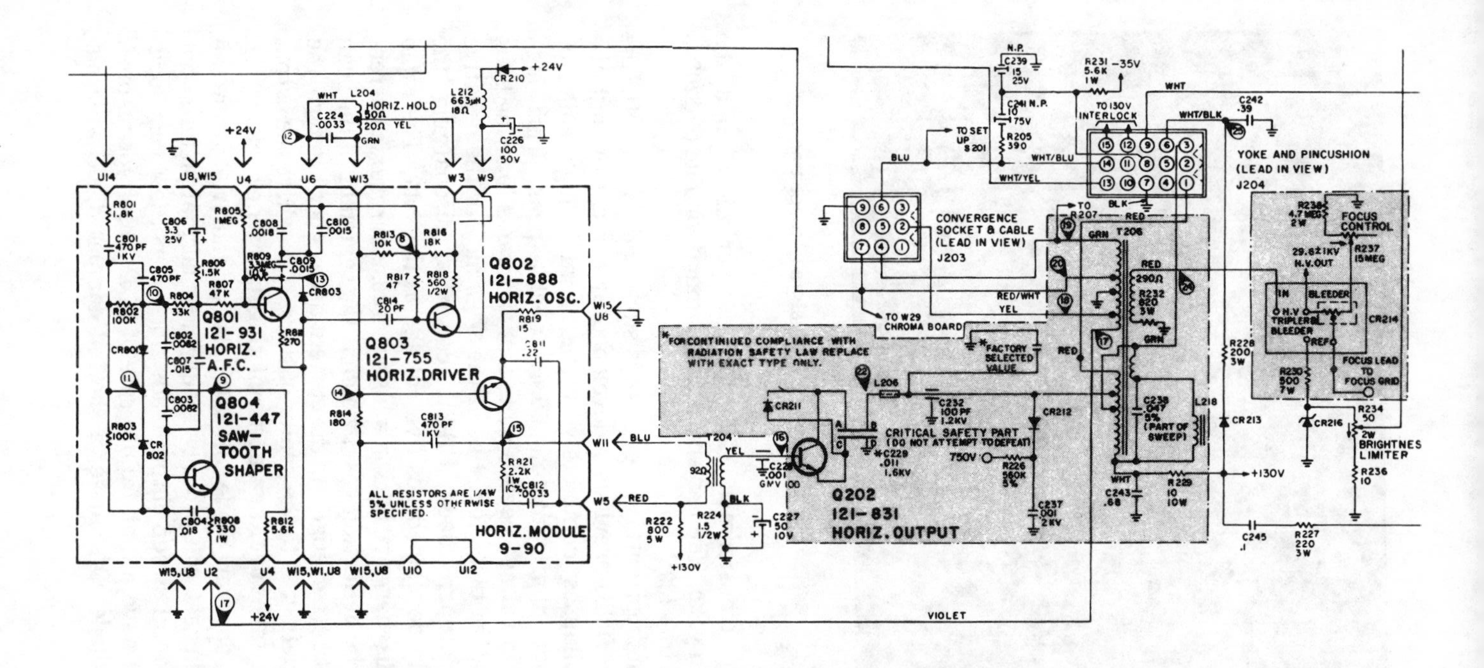

Fig. 4-17. Zenith horizontal module circuit.

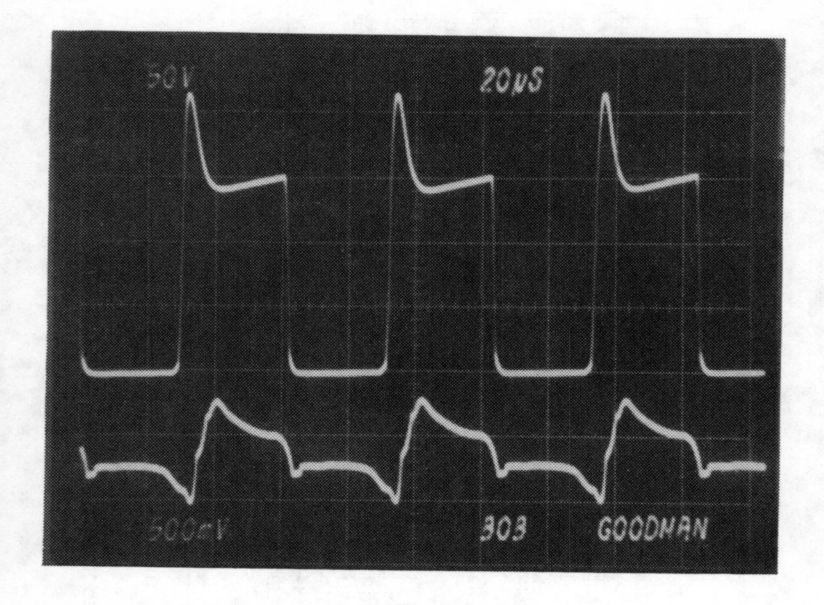

Fig. 4-18. Horizontal drive pulses (dual-trace scope trace).

of sweep transformer T206, the damper diode, and the deflection yoke. The low end of the primary winding of T206 is returned to +130V.

When horizontal output transistor Q202 is cut off, the magnetic field, previously developed by current through the yoke, collapses completely, resulting in a high collector voltage of approximately 1030V peak-to-peak. This high increase in voltage decays quite fast and wants to swing back in a negative direction. But at the beginning of the negative swing, damper diode CR-211 conducts, discharging the previously developed positive voltage at the yoke. The rapid voltage discharge results in a high increase in *reverse* current through the deflection yoke, which is the beginning of the scan at the left screen edge. As the current decays back to zero, the beam is deflected to the center of the raster, where the output transistor takes over again, completing the sweep cycle.

With damper diode CR-211 open or removed from the circuit, this set will produce a very near normal picture—but the horizontal sweep output transistor will not last very long. So, if you have a high failure rate of Q202, check for an open damper diode. Of course, a shorted damper diode will shut down horizontal sweep operation and kick open the breaker in two or three seconds.

In the older "E" line chassis, several "redundant" retrace capacitors were used as illustrated in Fig. 4-19. The capacitors are required to guard against excessive high voltage, in the event a single unit failed. For safety reasons, HEW rules forbid the use of a single "lumped value" capacitor with a single, common ground return. Several redundant value capacitors were spread out at widely separated locations around the horizontal sweep output circuit in order to conform with these safety regulations. The long wiring required for the redundant capacitors could, in some instances, produce an undesirable ringing effect at the left side of the screen.

Late model "E" and all "F" series chassis use a unique 4-lead retrace capacitor (Fig. 4-20) that was developed to meet HEW requirements. The 4-lead unit is mounted on the horizontal output transformer heat sink as shown in the Fig. 4-21. One feature of this new 4-lead capacitor is that should there be an open-mode failure (within the capacitor), the direct current path is removed from the horizontal output transistor and therefore shuts down the horizontal sweep and HV operation. This feature completely eliminates any possibility of "excessive" high voltage being developed that

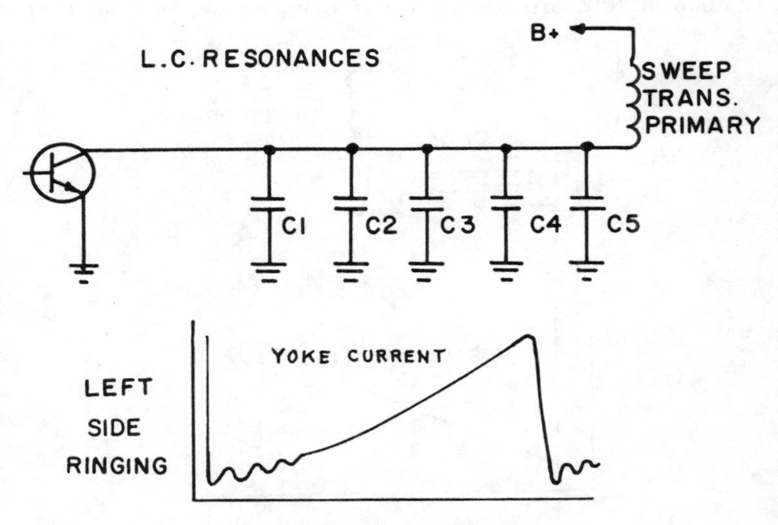

Fig. 4-19. Redundant capacitors.

could result from the open-mode failure of any one of the redundant capacitors used in the older model chassis. The compactness and short lead connections of this 4-lead capacitor also prevent the development of multiple resonance. Compare the 4-lead capacitor set-up in Fig. 4-20 with that in Fig. 4-19.

Should 4-lead capacitor C229 become shorted, you will have the same symptoms as with a shorted damper diode; that is, the circuit breaker will trip open in about two seconds. If the capacitor becomes leaky, the breaker may trip out after set is turned on from 5 to 20 seconds, or it may not trip at all, depending on the leakage resistance. There will also be no or very little high voltage developed. A good way to check for leakage in this capacitor is to operate the set for a minute or two and see if the capacitor is warm. If it is, the capacitor is defective.

If any lead inside the capacitor opens up, the horizontal sweep output stage will be shut down. Thus, no horizontal sweep or high voltage will be present. This is a good point to remember, since there has been a number of these 4-lead capacitors that have had open internal leads. This can also be an intermittent problem. To check, jump clip leads across terminals A and B, or C and D.

Driver transformer T204 and associated circuits can also develop a few problems. If the primary winding of T204

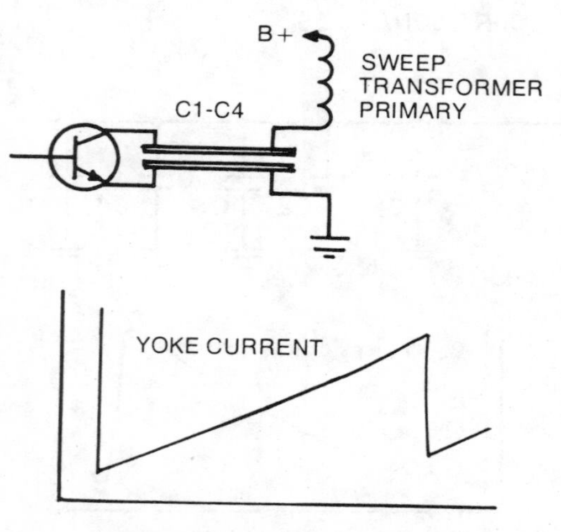

Fig. 4-20. Drawing of 4-lead capacitor.

Fig. 4-21. Location of 4-lead capacitor.

becomes partially shorted (an ohmmeter reading of about 40 ohms), driver transistor Q803 will become very hot and draw high current through R22, and this 800Ω resistor will burn out. If the horizontal oscillator stops, Q803 will also run very hot. But neither of these faults will likely trip the breaker.

If transformer T204 has internal arcs or a poor ground connection at its black lead, this can cause spikes to be fed into the base of Q202 and cause rapid failure of this transistor. And of course, an open R222 resistor will mean a loss of horizontal drive to base of Q202. These 5-watt resistors have been known to pen up intermittently. When R222 is open, the drive signal at pin W11 will look like the bottom scope trace of Fig. 4-18.

Probable Cause: The loss of sweep and high voltage in the cited case was caused by an open R222, an 800Ω 5W resistor. Also check T204 for correct winding resistance readings.

ZENITH 19EC45 AND OTHER CHASSIS WITH 9-90 HORIZONTAL MODULE

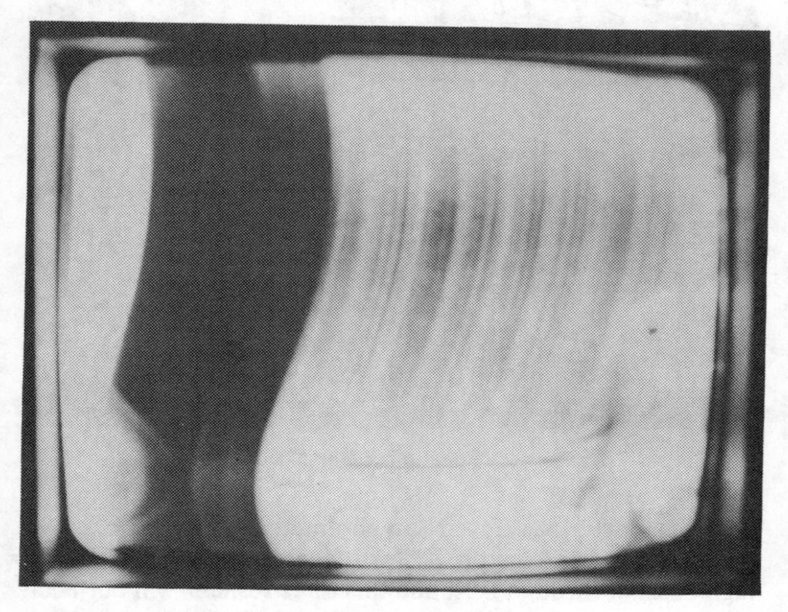

Picture Symptom: Picture floats across screen or has many slanted lines that indicate trouble in horizontal oscillator or sync circuits. Trouble could be within the 9-90 module or various pulses fed into this module. Because this was found to be an AFC fault, an analysis of this circuit would now be helpful.

Circuit Operation: Looking at the 9-90 module circuit in Fig. 4-22, you will see that negative-going sync is coupled to the cathode of APC diodes DR-801 and CR-802 through R801 and C801. At the same time, a negative-going pulse taken from the sweep transformer is fed to the emitter of Q804, via R808, to the sawtooth shaper transistor. Transistor Q804 is biased into saturation by the pulse, discharging C803. This very rapid discharge serves to sharpen the leading edge of a sawtooth waveform. The remaining portion of the sawtooth waveform is developed as C803 again charges through R8092. The resultant sawtooth waveform is coupled to the APC diode circuit through C802.

When the station sync pulse and the horizontal oscillator are exactly in phase, the APC comparator produces zero error voltage. Should the oscillator either lead or lag the sync pulses an "error" voltage is produced at the anode of CR801. This error voltage can be positive or negative and will either add or subtract from the existing positive voltage appearing at the base of the horizontal AFC transistor Q801. Capacitor C807 and RC network C806-R806 comprise the AFC anti-hunt network. Q801 thus acts as a variable resistance in series with C809.

The emitter is returned to ground through R811. The DC collector voltage is obtained by rectifying the 15.75 kHz sine

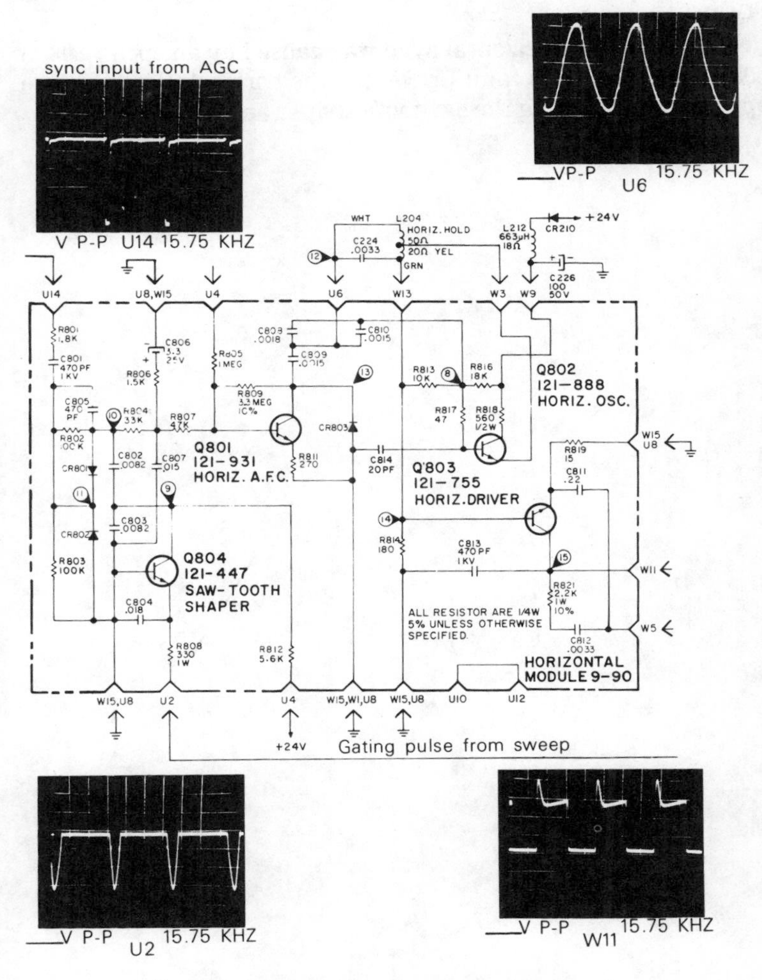

Fig. 4-22. Zenith 9-90 horizontal module circuit.

wave generated by the horizontal oscillator, which is coupled through capacitor C809 to the function of Q801 and the cathode of diode CR-803. On the negative portion of the sine wave, diode CR-803 conducts, resulting in a net positive charge on the collector side of C809, providing the required positive DC collector voltage.

Probable Cause: The first step for this type trouble is to check with a scope for proper horizontal sync pulse at pin U14 and a gating pulse at pin U2 of the module. Component suspects on the module would be CR-801, CR-802, Q804, and Q801.

This loss of horizontal sync was caused by an open R808, a 330Ω resistor. This fault prevented the horizontal comparison pulse from reaching the sawtooth shaper and AFC diodes.

ZENITH 19EC45 COLOR CHASSIS

Picture Symptom: No picture. Loss of high voltage. The screen is dark, but the set has good sound. Check for trouble in horizontal sweep output section or the 9-90 horizontal module.

Circuit Operation: Referring to the 9-90 horizontal module (Fig. 4-23), note that horizontal oscillator transistor Q802 is used as a modified Hartley circuit. This type oscillator has the parallel-tuned tank circuit connected between the base and collector of the bipolar transistor, and the inductive element of the tank coil has an intermediate tap that goes to the emitter. The resonant circuit consists of L204 and C224. The frequency of this tuned circuit is determined by the setting of coil L204, the horizontal hold control. The 15.75 kHz waveform generated by the oscillator is coupled to the base of horizontal driver transistor Q803.

Probable Cause: This picture symptom was caused by a loss of a horizontal drive pulse. A new 9-90 horizontal module was installed but the horizontal oscillator would still not operate. Note that the oscillator circuit (Fig. 4-23) has some outboard components that could cause some problems, so don't forget to check these.

A dead oscillator could be caused by an open L204 horizontal hold control coil or a shorted C224 capacitor across

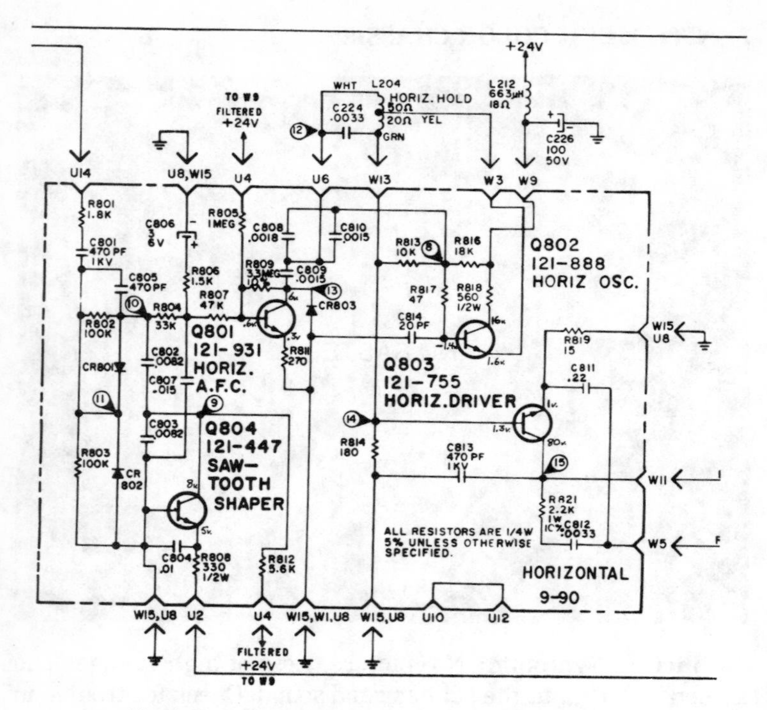

Fig. 4-23. Zenith 19EC45, horizontal sweep circuit.

this coil. Also look for loss of voltage to the oscillator stage. Check for +24 volts at pin W9 of the module. A loss of +24V at W9 was the trouble in this set—choke L212 had opened up. This could have been caused by a shorted C226 bypass capacitor or a faulty module.

When the horizontal oscillator is not working, a condition exists that may damage other module components. Driver transistor Q803 will be biased on, causing this transistor to become extremely hot, and this could give a false indication that the trouble is in the driver stage, when it is not.

Some Module Quick-Checks:

1. Take a few moments and make a thorough visual check of the module PC board for any obvious damaged or loose components.
2. Check or replace all discolored or charred resistors.
3. Test all solid-state devices: ICs, transistors, diodes, etc.

4. Repair any cracked PC boards or cold solder (loose) joints. Also note any component leads that may have shorted.
5. With the module plugged in and the set powered up, run through all the key DC voltage test points, namely the EBC of the transistors. This will help to quickly pin down the fault.

Picture Symptom: Picture appears to have horizontal-width (drive) and fold-over problems. A wide drive line may appear in center of screen. At times it may have a Christmas tree effect. An intermittent trouble may also damage horizontal output transistor Q202. Look for trouble in the horizontal sweep, interstage coupling system.

Circuit Operation: A square-wave signal is generated in the collector circuit of Q803 located on the 9-90 horizontal module (Fig. 4-24). This square wave is coupled to the base of horizontal output transistor Q202 via horizontal interstage driver transformer T204. As shown in Fig. 4-25, a 150V P-P square-wave should be found at module terminal W11.

The base of horizontal output transistor Q202 is connected to the top of the secondary winding of the horizontal interstage transformer. The bottom of this winding is returned to ground through resistor R224 and is bypassed by a 50 μF capacitor. Be on the lookout for a poor ground connection (cold joint) at this point. The emitter of Q202 is returned directly to ground. Ferrite bead L206 on the collector lead suppresses "spooks" generated by the rapid switching of the transistor junctions. The collector is connected to the primary of sweep transformer T206, the damper diode, and effectively to the

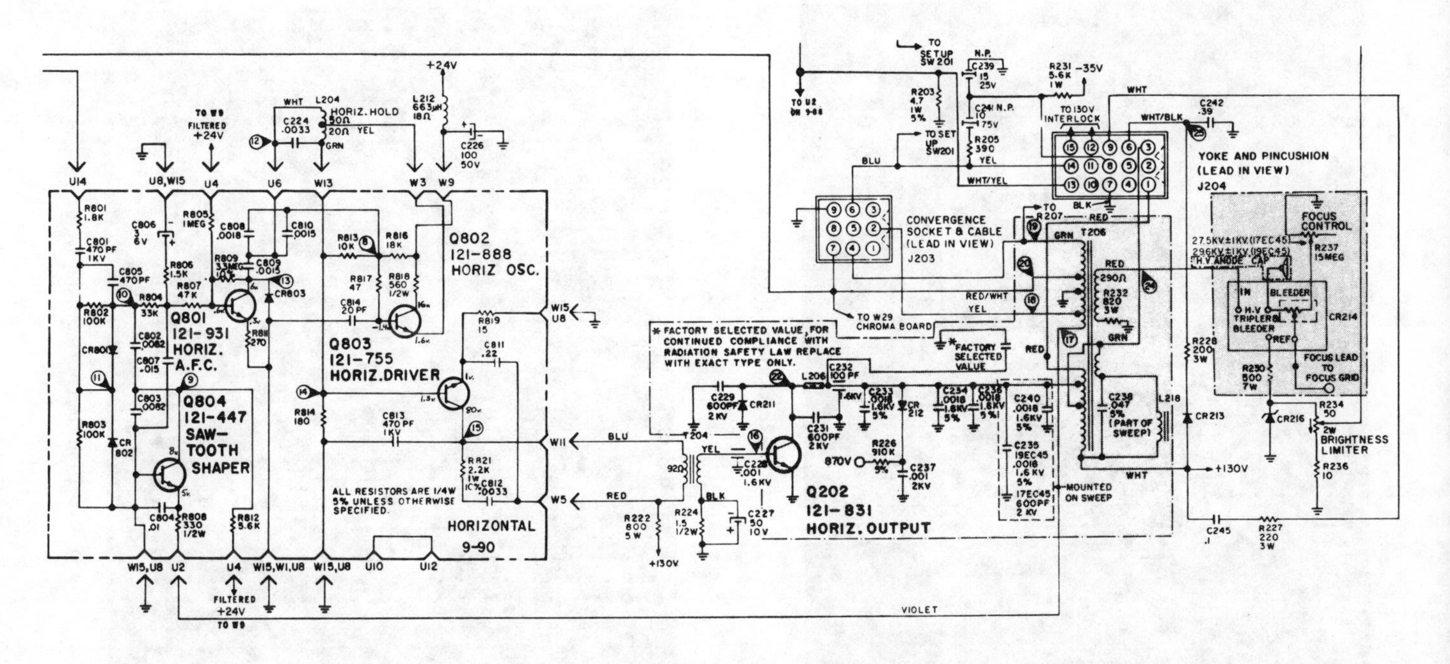

Fig. 4-24. Zenith sweep circuit.

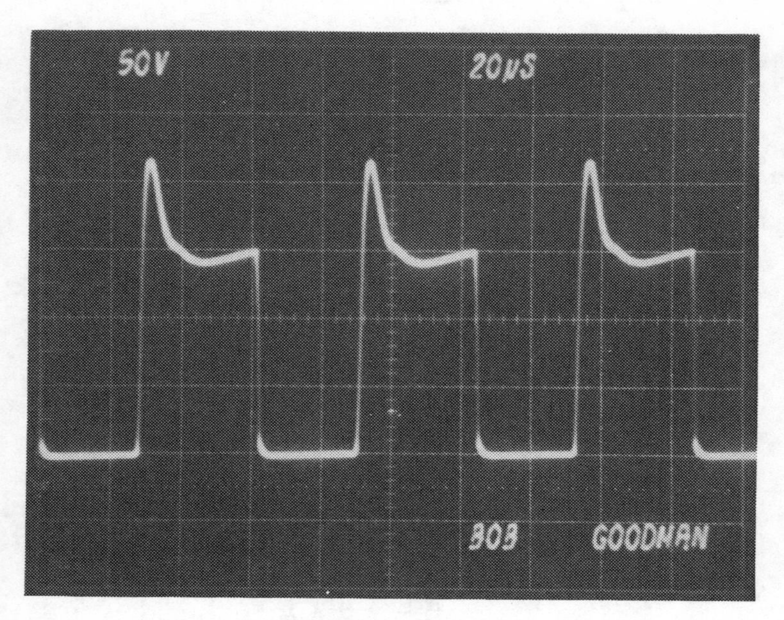

Fig. 4-25. Correct drive pulse.

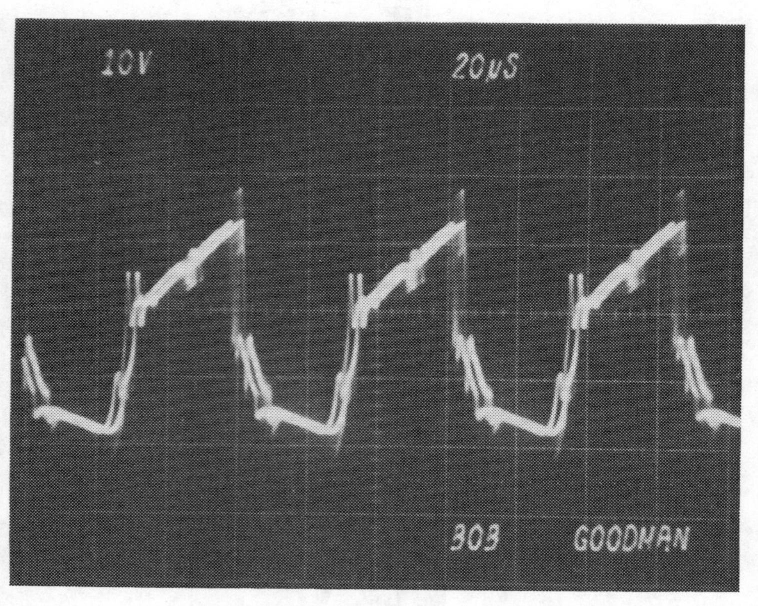

Fig. 4-26. Distorted horizontal drive pulse.

deflection yoke. The lower primary winding of T206 is returned to +130 volts.

Probable Cause: To isolate these narrow picture symptoms, first install a new 9-90 horizontal module. If this does not have any effect, use a scope to check the horizontal drive pulse at terminal W11 and at the base of Q202.

In this problem chassis, the drive pulse at module terminal W11 (primary of T204) was not correct. As shown in the Fig. 4-26, the amplitude was only 30V P-P and the pulses were distorted. Also note the high spikes that destroyed output transistor Q202. The low-amplitude pulse and loss of picture width was caused by shorted primary windings in T204, the interstage driver transformer. An ohmmeter check of the primary indicated only 48 ohms, while the replacement transformer checked out at 92 ohms.

Chapter 5

Vertical Symptoms

Of course, it's not possible to cover in a single book all problems that could exist, or all the cures for any conceivable problem. But the vertical symptoms shown here are representative of a great many receivers. When you come across a problem that is obviously related to the vertical circuits of a specific receiver, you can approach it in two ways:

1—Look at the TV trouble photos in this section for symptoms nearest to the appearance of your problem set. Even if the circuits are not the same, the basic theory will generally apply in most cases—unless some very perculiar circuitry is involved.

2—Look through the circuits presented in this section for those that come closest to approximating the one you are concerned with. Even if the problem is not the same, you should be able to get a good idea of the general operational theory, and thus be able to search out the culprit regardless of how different the circuit may be.

GENERAL ELECTRIC YA CHASSIS

Picture Symptom: Horizontal white line across screen. In most sets this would indicate a malfunction in the vertical sweep circuits or vertical module in later model solid-state chassis. However, in this chassis the problem could very well be found in the scan-rectification power system. Let's see how and why.

Circuit Operation: The scan-rectified voltage is developed from a winding on the horizontal sweep transformer. The scan rectifier diodes conduct during the long scan interval, as opposed to a boost diode which conducts during the retrace interval. This allows the use of smaller value filter capacitors than would be the case for a conventional 60 Hz, half-wave, rectifier circuit.

The circuit in Fig. 5-1 illustrates how 2 scan rectifier diodes are used to develop the positive and negative 13 volts used in the YA chassis for the vertical deflection stage.

Circuit Checks: An open condition, in the sweep transformer windings for example, would reduce one of the 13V supplies to zero, resulting in a vertical sweep problem. The results of shorted diodes, filters, transformer windings, or

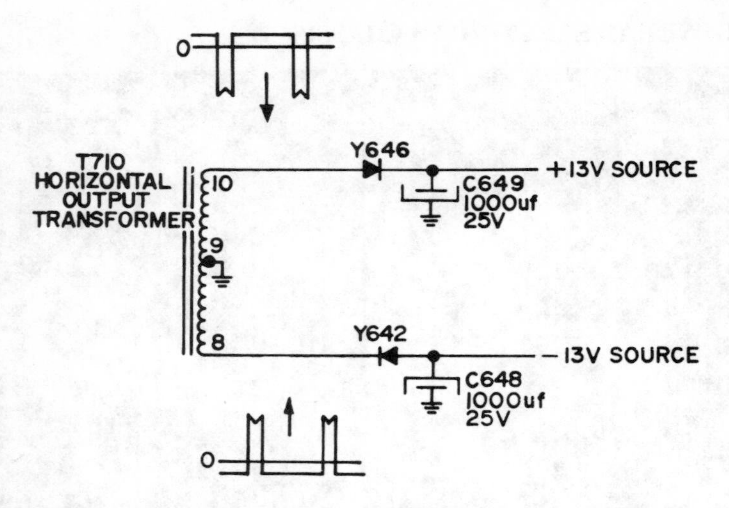

Fig. 5-1. G.E. scan-rectification circuit. A centertapped winding on the horizontal output transformer supplies positive and negative voltages for the vertical sweep module.

vertical module will be quite different. A high current load in this circuit would cause the B+ fuse to flow in the 60 Hz low-voltage power supply. The horizontal output transistor would demand more current, which in turn draws its current from the fused 60 Hz power supply. Shorted scan rectifier diodes Y646, Y642, or a short in the vertical module could burn out the scan winding on horizontal sweep transformer T710.

The rectifier diodes used in scan-rectifier circuits conduct about 80 percent of the time, and therefore must be of the long duty-cycle type. In addition, they must be able to withstand the fast rise time, high-amplitude retrace pulse. For this reason, exact replacement parts should always be used in these circuits.

Troubleshooting any TV chassis for a current overload is largely a process of elimination . The most obvious component faults would be filter capacitors, horizontal output transistor, damper diode, and wiring shorts. If the cause of overload is not located, then disconnect the scan-rectified B+ supplies, one at a time, while quickly monitoring the 60 Hz power supply's B+ current with an ammeter. Current will return to normal when the defective stage is disconnected from the horizontal output transformer. The scan rectified circuit in this chassis can be disconnected from the sweep transfomer by removing the vertical plug-in module but be sure to reduce the brightness level to prevent burning a line across the screen of CRT.

Probable Cause: This trouble was caused by shorted output transistors in the vertical module, which also burned up the winding on the horizontal sweep transformer. To correct this problem, the sweep transformer and vertical module were replaced.

Picture Symptom: This white horizontal line across screen. No vertical deflection. Look for trouble in the vertical deflection VA panel or deflection yoke.

Circuit Operation: The VA panel (Fig. 5-2) contains the vertical and horizontal sync circuits, as well as the vertical sweep stages. Thus, do not overlook this panel should any sync problems arise.

The low-frequency video signals enter the VA panel on point 11 and are coupled to the sync separator. The output is applied to both the vertical oscillator and the inverter stage. The vertical oscillator is synchronized with the station signal by vertical sync pulses. Horizontal sync is inverted by Q6 and fed via pin 10 to the FC horizontal sweep panel.

The vertical blocking oscillator conducts and discharges capacitor C5 during retrace time. During vertical scan time a sawtooth wave is formed across this capacitor and is amplified by the pre-driver, driver, and complementary output stages. This output leaves the panel at pin 2 in order to drive the yoke, pin cushion, and convergence circuits. Top and bottom raster correction is obtained through feedback circuits to the predriver.

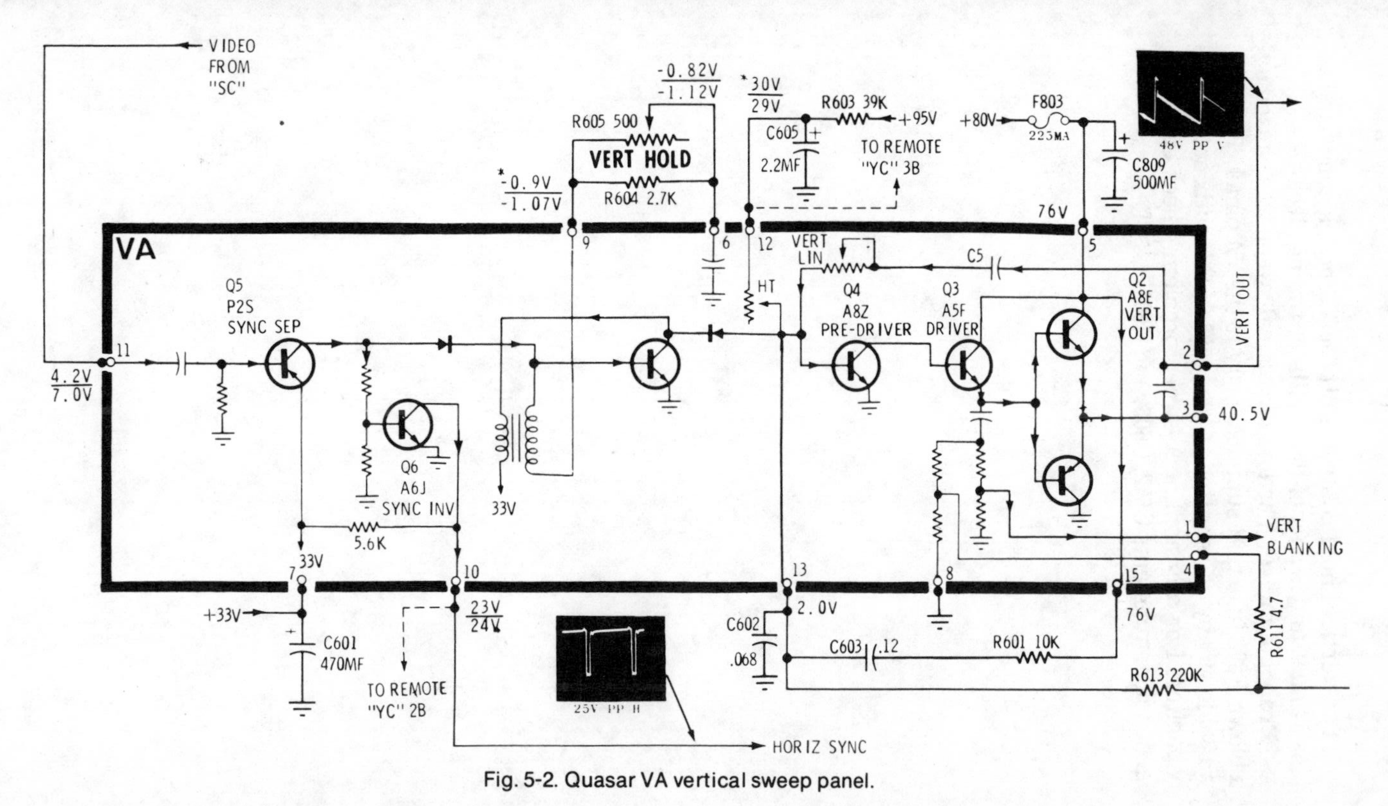

Fig. 5-2. Quasar VA vertical sweep panel.

The vertical hold control is off the VA panel, connected to pins 6 and 9, and is part of the oscillator timing circuit.

Probable Cause: The loss of vertical sweep was caused by a blown F803 fuse that supplies +80V to the VA panel. The fuse was blown by a shorted C809 (500 μF) outboard filter capacitor. For any nonlinear sweep problem, check outboard feedback components C602, C603, and R601.

Picture Symptom: Picture wants to roll. Can be stopped with the vertical hold control, but will start rolling again after set warmup. The picture *will* lock in, which indicates no trouble in the sync circuits. Look for trouble in the vertical oscillator or feedback circuits.

Circuit Operation: Refer to the vertical circuit in Fig. 5-3. Amplified vertical sync pulses are obtained from the plate (pin 8) of the 6HS8 sync-AGC amplifier. These are negative-going, 60V peak-to-peak pulses. Any horizontal pulses will be filtered out by the 68K resistor and 0.01 μF capacitor. After passing through the integrator A1, a 3V P-P negative-going pulse appears across the diode X5 in the cathode circuit of vertical oscillator V5A. Diode polarity is such that it acts as a load for the sync pulse, yet appears as a short circuit for the current in the cathode circuit.

During conduction, the grid of vertical oscillator V5A develops a 95V P-P pulse that is amplified through the plate, shaped by C76 (0.1 μF) and the 1K resistor to a sawtooth voltage waveform of 70V P-P, and delivered to the grid of the vertical amplifier V5B. The vertical output plate pulse is a spike of 1400V P-P amplitude, which has a slight sawtooth effect on the base line. Signals from the vertical output

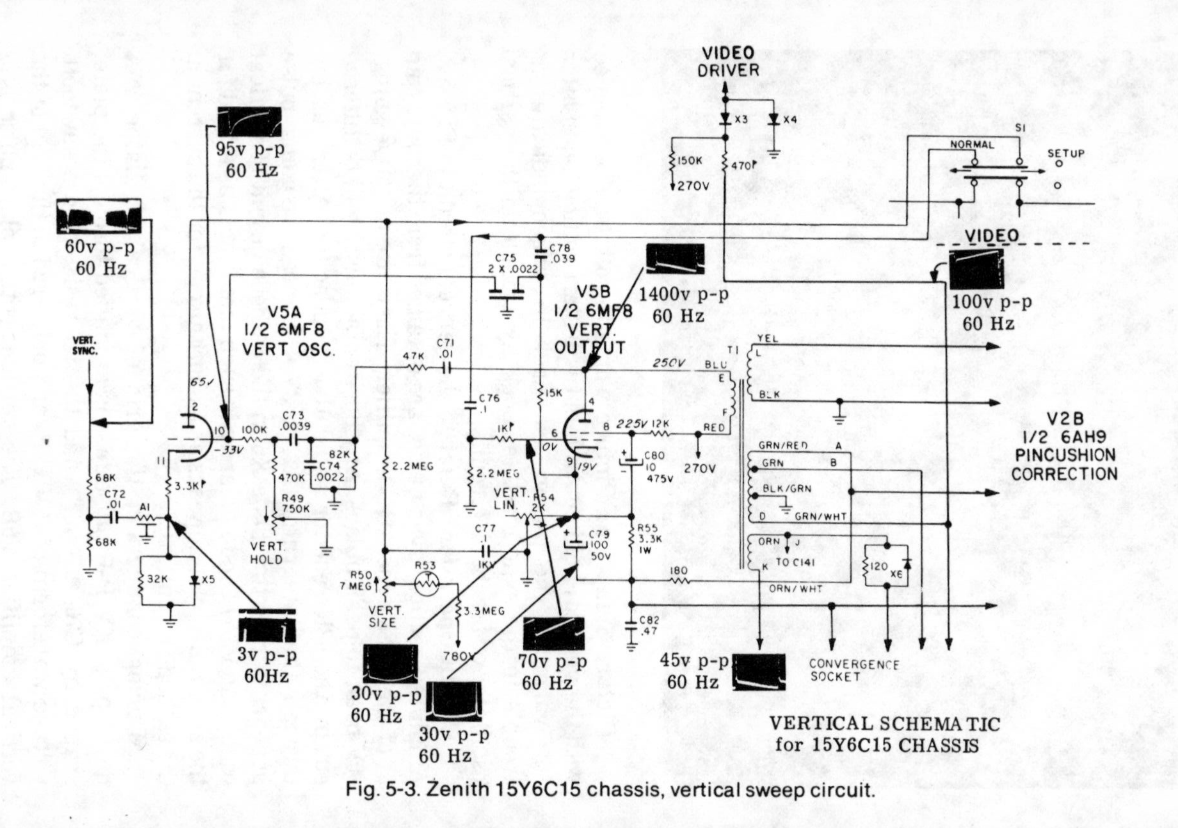

Fig. 5-3. Zenith 15Y6C15 chassis, vertical sweep circuit.

transformer are used for vertical sweep, pincushion correction, convergence, and blanking. Part of this vertical spike is attenuated and shaped in the feedback loop via C71 the 47K and 82K resistors, C74 and C73. The resulting 540V spike at the 100K resistor leading to the grid of the vertical oscillator is a trigger to fire the vertical oscillator again and create the next vertical scan.

The vertical hold control in the grid circuit of the oscillator controls the triggering of the oscillator itself on the feedback pulse, allowing the low-amplitude sync pulse in the cathode circuit to maintain lock-in. The vertical size control affects the plate voltage and thus the amplitude of the oscillator. The linearity control is in the cathode circuit of the output stage and is used to control the waveshape of the output stage.

Probable Cause: This vertical roll symptom was caused by leakage in C73, a 0.0039 μF capacitor. Any of the components in the feedback circuit such as C71, C74, and the 47K or 82K resistors can cause a vertical roll problem. If diode X5 shorts, the picture will not lock due to loss of the sync pulse. If X5 opens, the picture height will be reduced, and it may also roll.

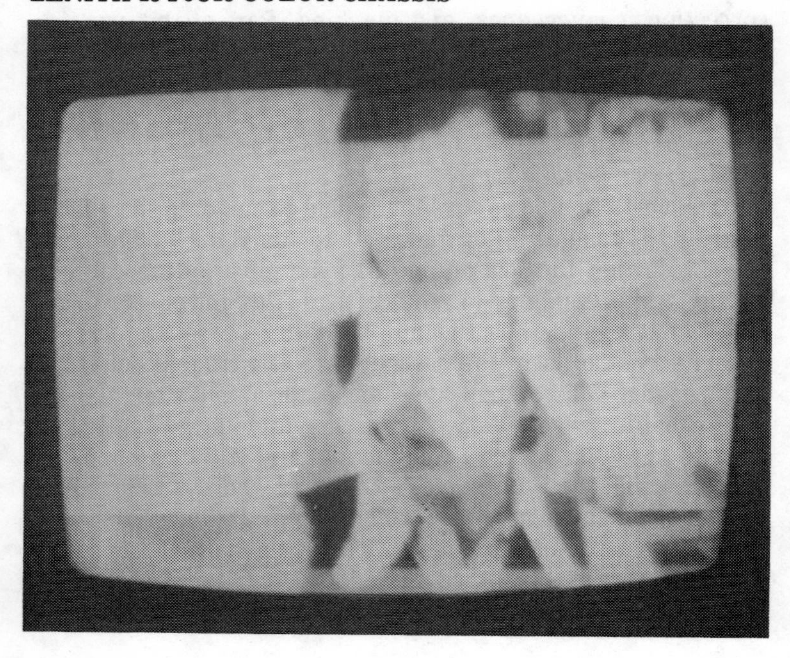

Picture Symptom: Picture had a vertical roll. The picture could not be stopped with the vertical hold control, but there was some effect on the speed of roll. It appeared the vertical oscillator was running at twice the normal vertical oscillator rate.

Circuit Operation: This color chassis uses a common common vertical sweep system as shown in Fig. 5-4. An oscillator and amplifier is employed, with the amplifier acting as part of the oscillator.

The feedback circuit for the vertical deflection can be easily seen. Vertical oscillator triode V5A discharges and couples a strong voltage pulse into the vertical output grid. V5B is a class-A amplifier coupling a higher pulse to the output transformer and yoke for deflection. A portion of the output pulse is looped back from plate pin 4 through C71 and a 47K resistor, then on to C73 and a 100K resistor, to the oscillator grid (pin 10) to trigger the next discharge pulse. The vertical sync pulse is coupled into the cathode (pin 11) at a point above diode X5.

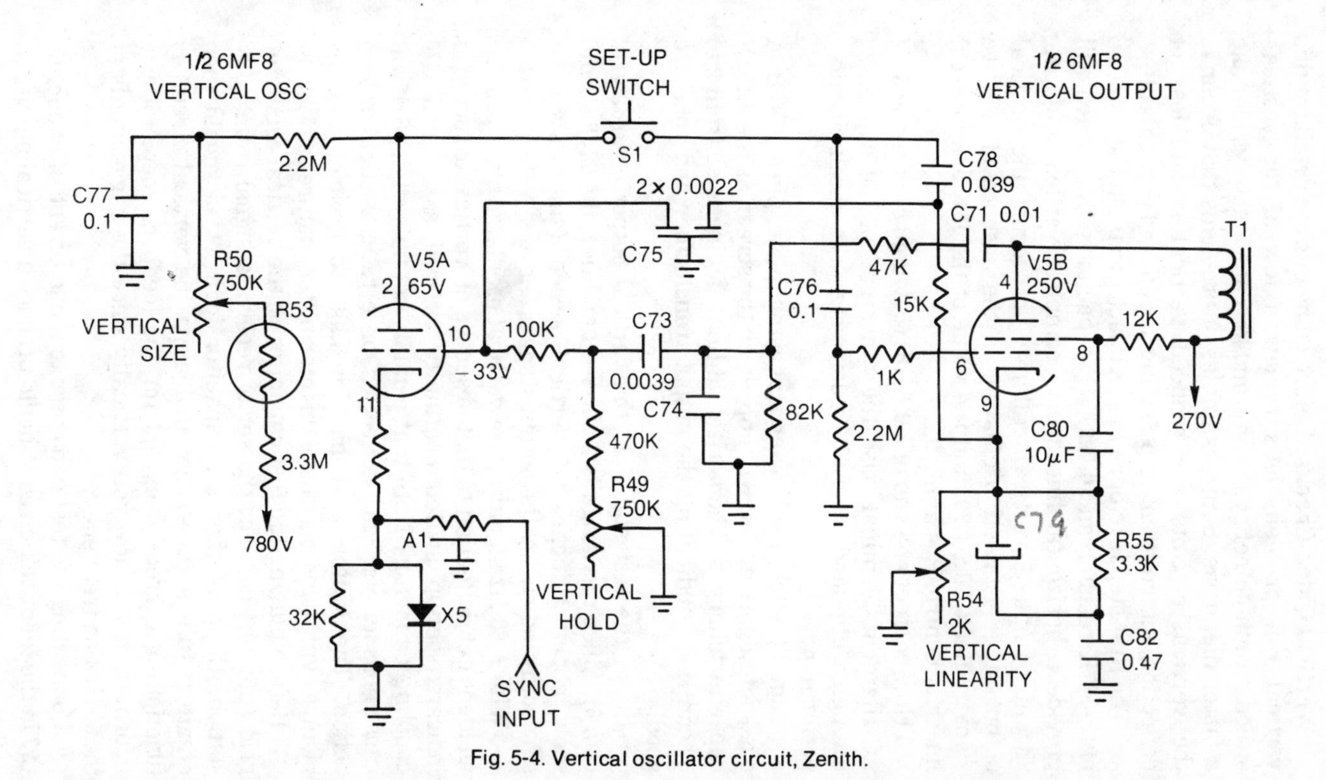

Fig. 5-4. Vertical oscillator circuit, Zenith.

Vertical Quick-Checks: Before going into the details of this vertical roll problem let's review a few of the vertical-sweep-circuit defect symptoms and look at some ways to make a faster diagnosis. Some examples of problems that occur in the vertical circuits are reduced vertical sweep, loss of deflection, intermittent or pulsating sweep, and vertical roll.

The first step is to open the feedback circuit at the junction of the 47K resistor and 0.0039 μF capacitor (C73), which will provide a horizontal line due to complete vertical collapse. Should complete vertical collapse *not* occur, the vertical output stage is itself oscillating in the absence of a drive signal. Check C82, C79, and the value of the 15K and vertical linearity control R54 at pin 9.

Now correct one end of a 0.1 μF, 600V capacitor to the 6.3V AC filament terminal. Use the other end of the capacitor, through a clip lead, as a 60 Hz drive signal to check circuitry performance.

Touch the test lead to vertical output plate, pin 4 of V5B. A one-shot negative kick will appear on the screen as the 0.1 μF capacitor charges. The horizontal line may increase from 2 to 3 scan lines, indicating the output transformer and yoke are good.

Inject the test signal at the vertical output control grid (pin 6). Since this stage is a class-A amplifier, the CRT should show about 3 inches of vertical sweep (sine wave, not sawtooth) indicating the output stage is okay.

With a 60 Hz signal at the vertical oscillator plate (pin 2), an identical deflection to that observed on vertical output grid indicates set-up switch continuity as well as a good C76 coupling capacitor. Note that a faulty set-up switch (S1) will cause loss of vertical deflection or erratic vertical sweep action. The switch may have leakage to ground, and the picture symptoms may be of an intermittent nature.

Before placing the test capacitor on the vertical oscillator grid (pin 10), connect the set to receive a signal from a TV station. Clipping the test capacitor to point 10 should then create a full picture scan, indicating the vertical circuit is functioning correctly, except for the feedback loop. Picture information will now roll vertically since the triggering is at the 60 Hz line frequency.

Connecting the test capacitor to the loose end of capacitor C73 in the feedback circuit should produce a vertical picture at a reduced amplitude. It is also possible to take hold of C73 with the fingers and obtain some vertical sweep, with the amplitude being dependent upon the amount of 60 Hz picked up by your body capacitance.

If diode X5 in the cathode of V5A shorts out, there will be no vertical picture sync lock—vertical sync pulses will be shorted to ground. An open X5 will reduce the vertical sweep.

Probable Cause: This vertical roll was caused by leakage in C71, a 0.01 μF feedback capacitor from the plate of V5B.

Picture Bending or Pulling—Loss of Horizontal and Vertical Lock

Many times you do the routine checks for these symptoms, only to find that there is no defect in the component you thought would cure the trouble. Picture weaving, for example, often results from poor filtering in the power supply, zener oscillation, line voltage phase problems, and combinations of problems that can routinely be fixed with a new electrolytic (sometimes of higher value) in the B+ supply. But the weaving often persists—perhaps intermittently, which can be a real mind bender—regardless of what you do. As many of these case histories presented here will prove, sync and AGC circuits can be, and often are, the culprits.

As you read through these case history troubles in an attempt to find a particularly persistent problem, you will notice that there are times when the problem itself is actually a part of the symptom. You may find a defective transistor, as an example, and its replacement may cure the trouble—*temporarily*. But a more serious, underlying problem has caused the transistor to fail. So when you locate the trouble and correct it, be sure to consider the possibility of some other fault causing the trouble in the first place. We often dismiss a transistor as being leaky, a resistor as being off value or open, and an IC that is inoperative. We may need to ask why. If nothing else, it will result in fewer callbacks, higher profits (we tend to absorb those callback costs), and more satisfied customers.

Picture Symptom: Complete loss of vertical and horizontal picture lock. Look for trouble in the sync separation or noise gate circuits.

Circuit Operation: For sync separation, NPN transistor Q302 is used in the sync clipper stage circuit (Fig. 6-1). Q300 provides buffering (isolation) and impedance matching to the input of Q302. Composite video (positive-going sync) is coupled to the base of Q302 via its signal-biasing network.

The positive sync tips forward-bias D303 and drive Q302 into conduction. This action charges up capacitor C308 to −0.6V at the base end. The result is an amplified negative-going sync signal at the collector.

During the time periods between sync-tip information, C308 slowly discharges through R316 and reverse-biases the transistor. As a result, only amplified and inverted sync tips of the composite video signal are seen at the collector. The amplified sync is then coupled to the horizontal and vertical sweep circuits.

The noise gate circuit's function is (Fig. 6-2) to provide noise immunity for the sync and AGC systems. During the

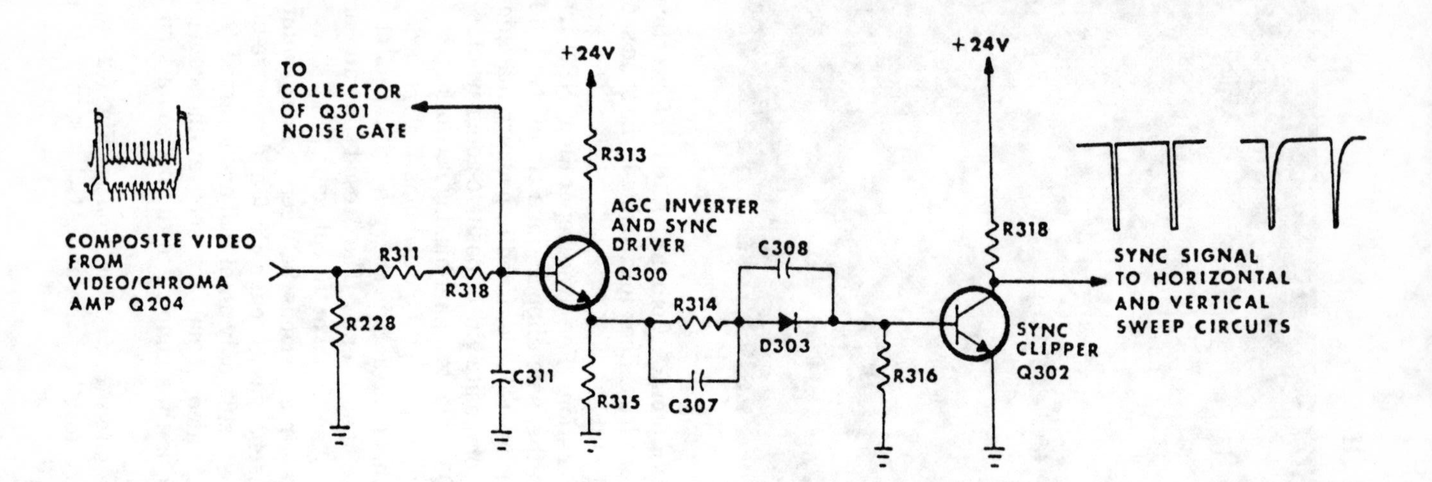

Fig. 6-1. Admiral M10, sync separation circuit.

presence of noise spikes, Q301 conducts, removing the input signal from Q300, the AGC inverter and sync driver. This prevents false triggering of sweep circuits because of poor sync separation and improper AGC action.

The base circuit of Q301 constitutes a "floating bias" arrangement that maintains proper relationship between base signal and DC bias during changes of input signal level. Rectifier D302 provides this action by becoming forward biased when the positive amplitude of an input signal exceeds 0.6V above peak sync level.

The remaining signal is the 0.6V peak-to-peak sync tips, which are coupled via C306 to the base of Q301. Unwanted noise spikes of greater amplitude than the sync tips will forward-bias Q301, whose collector is connected to the base input of Q300.

During Q301 conduction, the base of Q300 approaches ground potential, removing the noise spike and providing clean sync and AGC action.

Probable Cause: This loss of sync was caused by an open emitter resistor—R313 of Q300, the sync driver and AGC inverter transistor. Other possible faults could be a shorted Q301, open C307, shorted D303 diode, and of course a defective Q302 sync clipper.

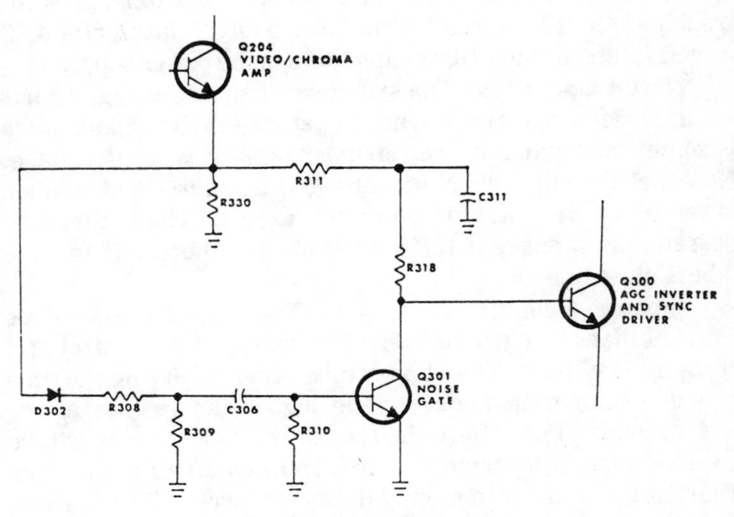

Fig. 6-2. Noise gate circuit.

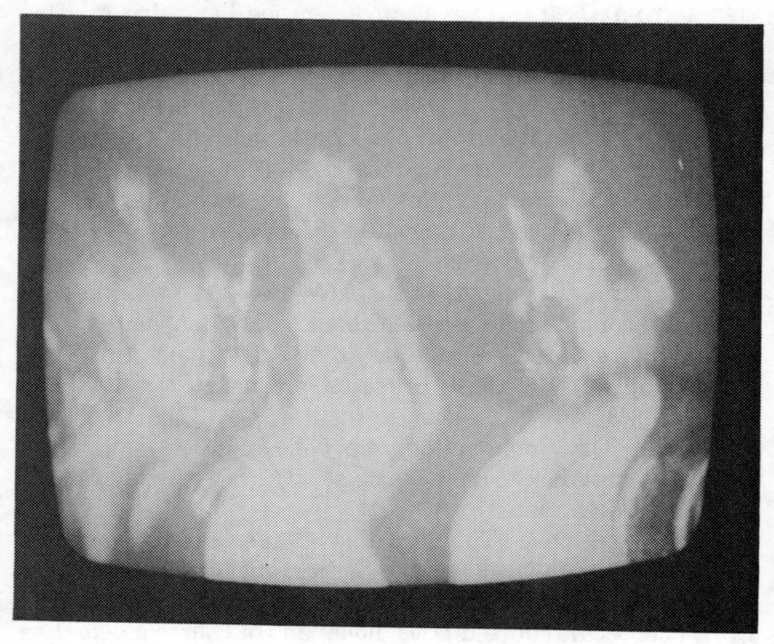

Picture Symptom: Horizontal bending or pulling. Look for trouble in the sync circuit. Could also be AGC, horizontal AFC circuit faults, or poor filter capacitors in the power supply.

Circuit Operation: The sync circuit shown in Fig. 6-3 has an amplifier stage and a sync clipper stage. The clipper stage is sometimes called a *sync stripper* or *sync separator* stage. Note that the clipper's plate voltage is about 95V and is much lower than you would generally expect. Thus, the tube operates as a sharp-cutoff amplifier, and the cutoff for this tube is close to zero.

A positive-going video signal (Fig. 6-4) that should be found at plate and grid of V500A is required at the control grid of the sync clipper. The clipper tube is cut off during the time of video information, due to the high negative bias at the control grid (pin 6). Only the tops of the sync pulsses will be positive enough to drive the tube from cutoff to conduction. Thus, at the plate output, only the sync pulses will be present. This then completes the process of clipping the sync pulses from the video signal information.

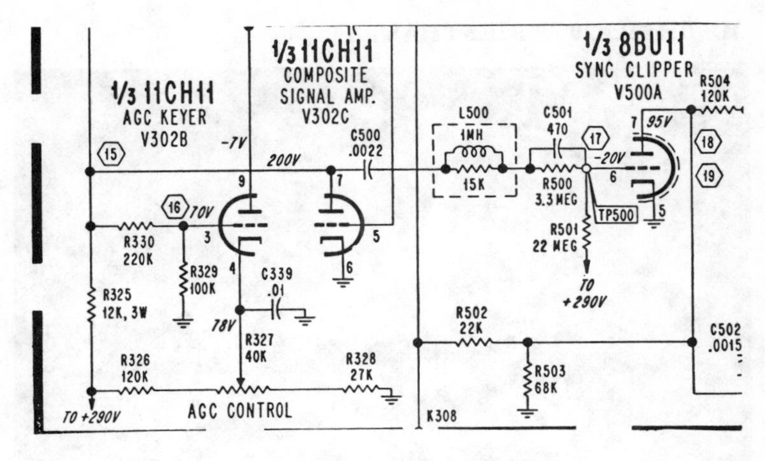

Fig. 6-3. Sync and clipper circuit.

Probable Cause: The horizontal picture bending in this set was due to some video getting into the sync pulses. To check, observe the waveform at the plate (pin 7) of the sync clipper stage. If any video information appears with the sync pulse, then coupling capacitor C500 is probably defective. In the cited case, C500 was leaky and a new 0.0022 μF unit returned picture to normal reception. A scope is a must for troubleshooting these sync circuits.

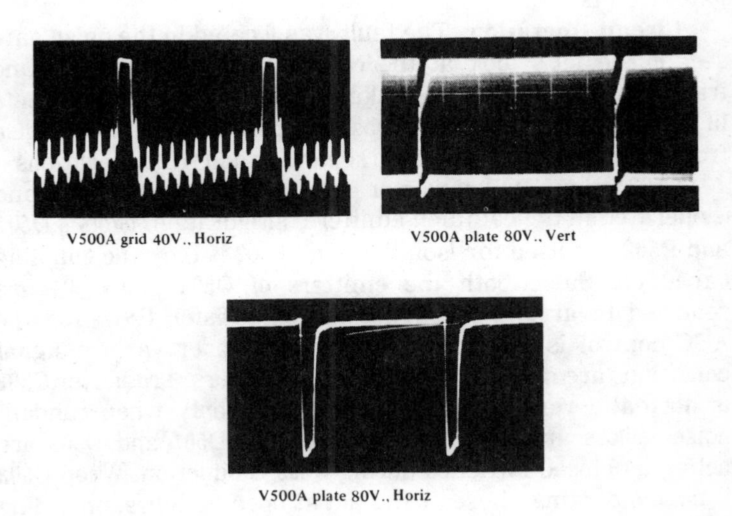

V500A grid 40V., Horiz

V500A plate 80V., Vert

V500A plate 80V., Horiz

Fig. 6-4. Scope waveforms of sync pulses.

Picture Symptom: Loss of vertical and horizontal sync. Look for trouble in the sync amplifier, sync separator, or noise gate circuits.

Circuit Operation: The fault was located in the noise gate section, so let's look at the circuit shown in Fig. 6-5. One transistor and two diodes make up the solid-state components in the noise-gate circuit, whose purpose is to prevent noise from interfering with the sync separator and AGC operations.

Q503 is an NPN transistor functioning as a noise gate and is operated in the common-emitter configuration. Diodes D501 and D502 are used for isolation, and C502 is the sync coupling capacitor. Since both the emitters of Q503 and Q504 are returned to ground through common resistor R512 and the AGC control is set, the operating bias under various signal conditions are maintained for both transistors. Transistor Q503 is normally reverse biased, conducting only when random noise spikes appear at its base. Diodes D501 and D502 are acting as closed switches during Q503 conduction. When Q503 is not conducting, these diodes act as open switches, providing isolation.

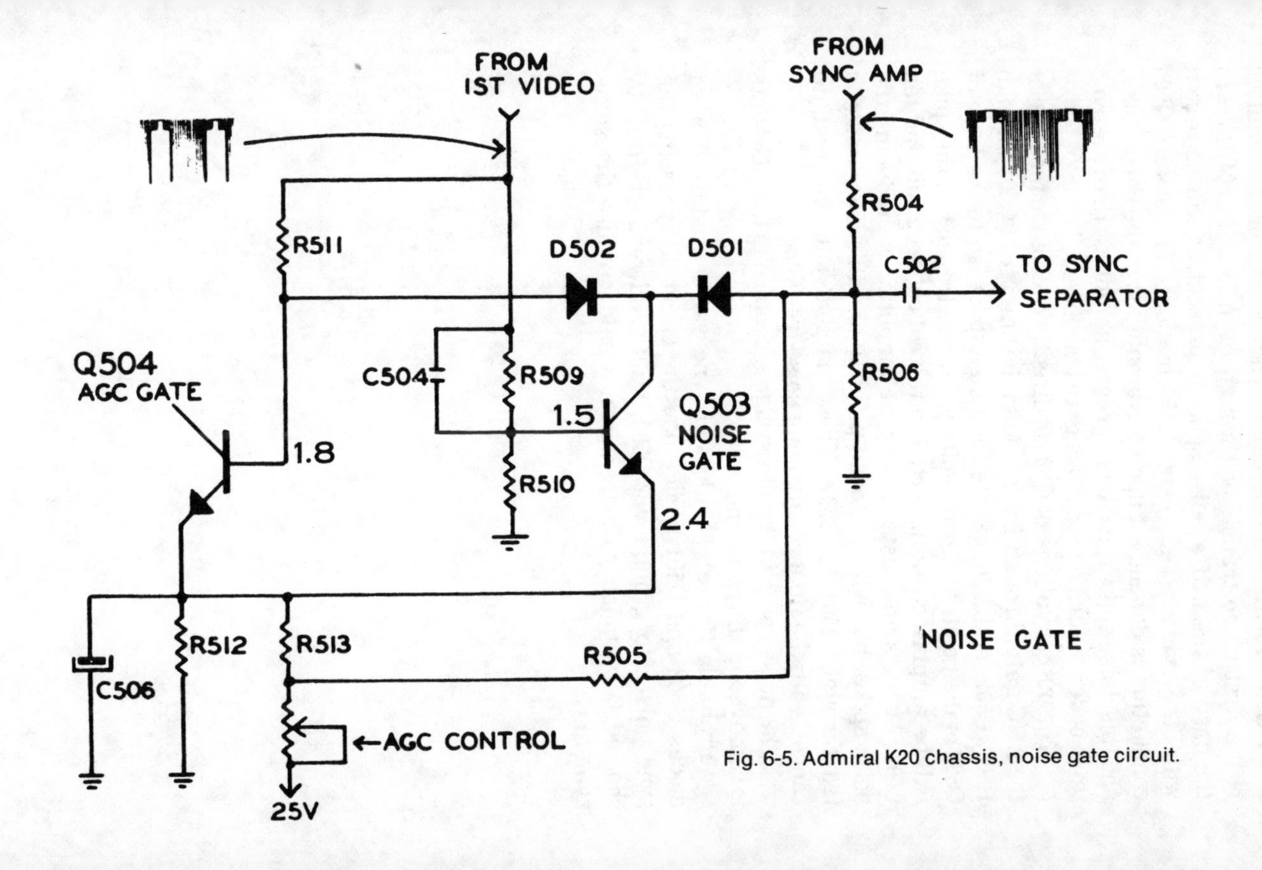

Fig. 6-5. Admiral K20 chassis, noise gate circuit.

Video information (including sync) is fed from the emitter of the first video transistor through RC network C504-R509 to the base of Q503. If a noise spike is present, it will be mixed with the video and sync information. Transistor Q503 conduction is dependent upon the amplitude of the noise spike, and this noise spike must be of greater amplitude than the sync tip in order to turn on the noise gate transistor.

The output voltage of Q503 is directly coupled to the base of the AGC gate through D502, and to the emitter of Q503 through R504, C502, and D501. Since the noise spike fed to the base of Q503 is the positive turnon voltage for this transistor, the spike will be amplified and inverted in the collector circuit and fed C502. At the same instant the sync amplifier is producing the noise spike, but with opposite polarity and this noise spike is fed through R504 to C502. These outputs join and tend to cancel, with the final result being noise-free sync.

The noise spike in the video signal coupled to the AGC gate is removed in the same manner.

Probable Cause: This loss of sync was caused by shorted diodes D502 and D501, which had the effect of bypassing the sync pulses to ground via Q503, Q504, and resistor R512. A shorted Q503 noise gate transistor would produce the same symptom.

MOTOROLA TS-941 CHASSIS (QUASAR)

Picture Symptom: Small, distorted picture with low distorted volume. Some chassis with same component fault may have no sound or picture. Look for trouble in the power supply circuits and ZC panel.

Circuit Operation: Refer to the schematic of the 80-volt regulated power supply in Fig. 6-6. Circuits of the AC panel convert AC line voltage to regulated 80V DC. This is accomplished by diodes and SCRs in a bridge circuit. The SCRs are pulsed by diac D6.

Initially, voltage from the start circuit charges C11. This causes diac D6 to conduct and forms a pulse across R12, sufficient in amplitude to trigger the SCRs. Once in operation, voltage for firing the diac is obtained from the DC output. At this time, Q5 conducts to isolate the start circuit from C11. The output also has provision for regulation of Q4.

Synchronization is obtained by controlling the starting time of the pulse generator during each alternation of current. This is accomplished by conduction of Q3 each time the bridge conducts, thereby removing the charge on the C11. At the end of each half cycle, Q3 turns off and C11 again charges to generate the next pulse from diac D6.

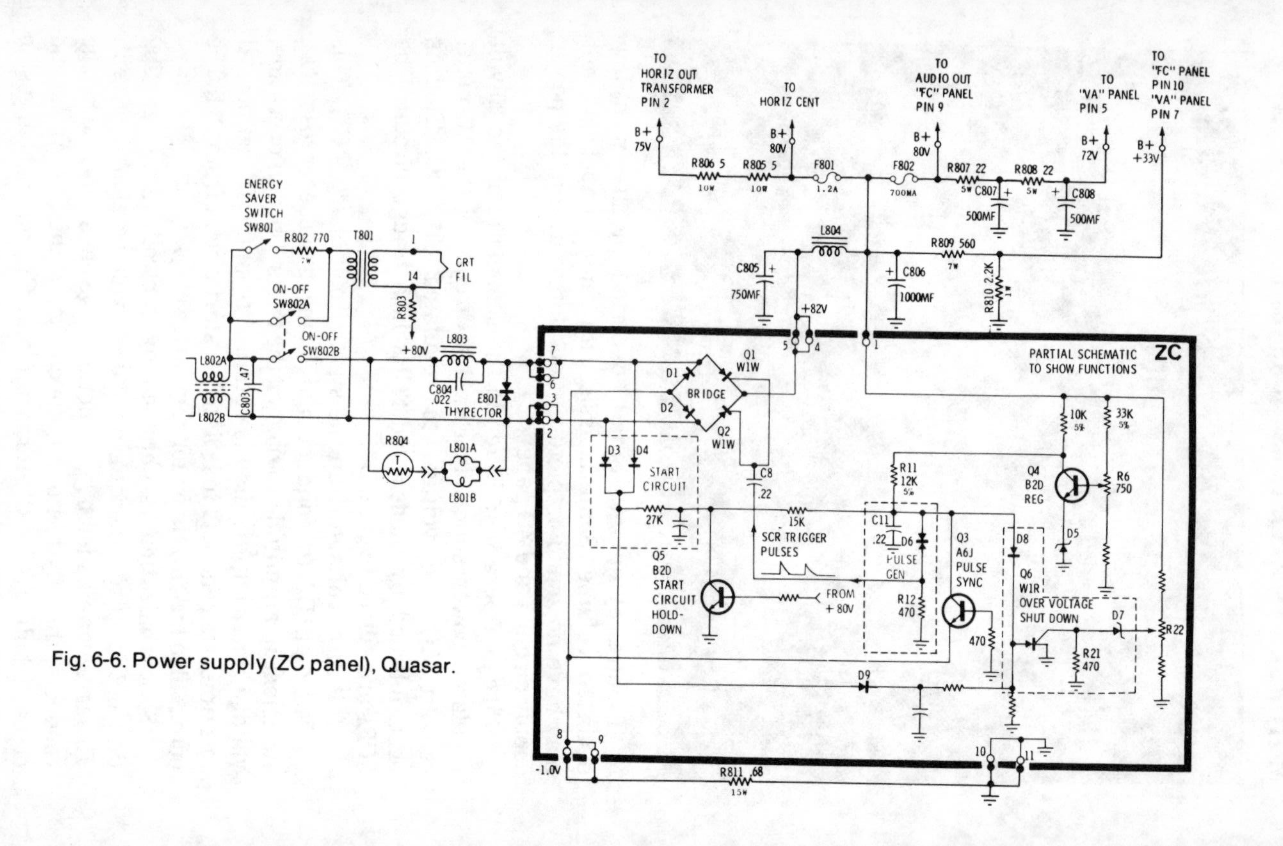

Fig. 6-6. Power supply (ZC panel), Quasar.

An overvoltage circuit shuts down the pulse generator if the output should increase beyond a safe level due to a defect on the panel. This provides protection against excessive high voltage and component damage. The supply is protected from heavy current overloads by a circuit breaker in the AC input circuit and by fuses in the supply lines. Thyrector E801 protects the power supply from any spikes that may come in on the AC line.

Voltage regulation is accomplished by controlling the duration of the SCR trigger pulses, which in turn control the *on* time of the SCRs located in the bridge rectifier circuit.

Probable Cause: This set's malfunction was caused by an increase in the value of R811, a 15W 0.68Ω resistor located on the main chassis. This is in the ground return of the bridge rectifiers, and the picture symptoms will vary with the change in value of R811.

Picture Symptom: Loss of horizontal hold and some AGC overload. Look for trouble in the AFC and 31.5 kHz clock system. Replace IC400 with a known good chip.

Circuit Operation: This chassis does not have a vertical or horizontal *customer control* adjustment because of the sophisitcated IC400 sync system. The noise protection, sync separation, and phase comparison in this chip make possible stable phase and drift-free frequency lock.

The 31.5 kHz clock is phase locked to the horizontal sync signal by a comparator circuit (note block D of Fig. 6-7). The 15.750 kHz sawtooth waveform and the horizontal sync are phase compared. Their phase comparison controls the clock timing or phase by speeding up, or slowing down, the clock's frequency (block E) in relation to the sync/sawtooth phase error.

Referring to the partial schematic of IC400 in Fig. 6-8, you will note that keying pulses from the horizontal sweep transformer are clamped to ground by diode SC425 and then fed to pin 8 of the chip. Q20 amplifies the pulse, inverting its polarity. Zener ZD1 clamps the signal and feeds the pulse to

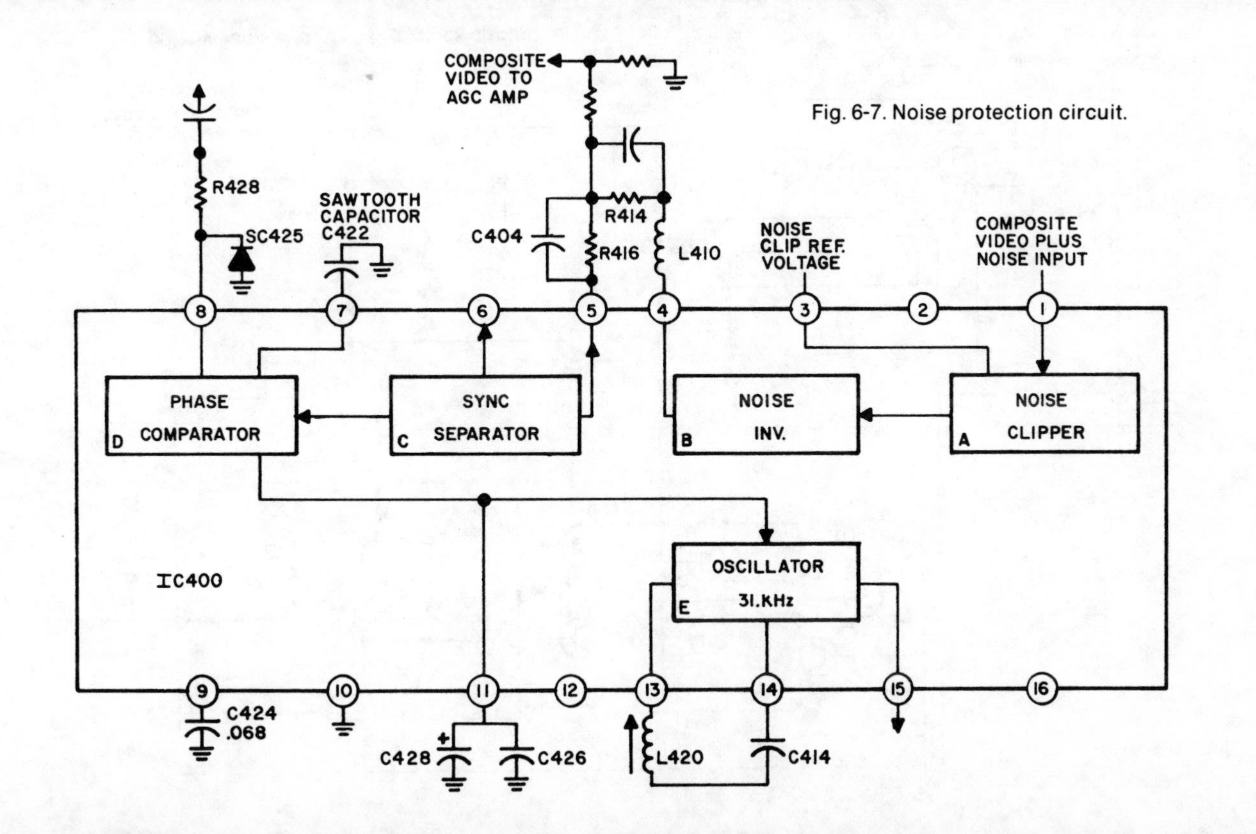

Fig. 6-7. Noise protection circuit.

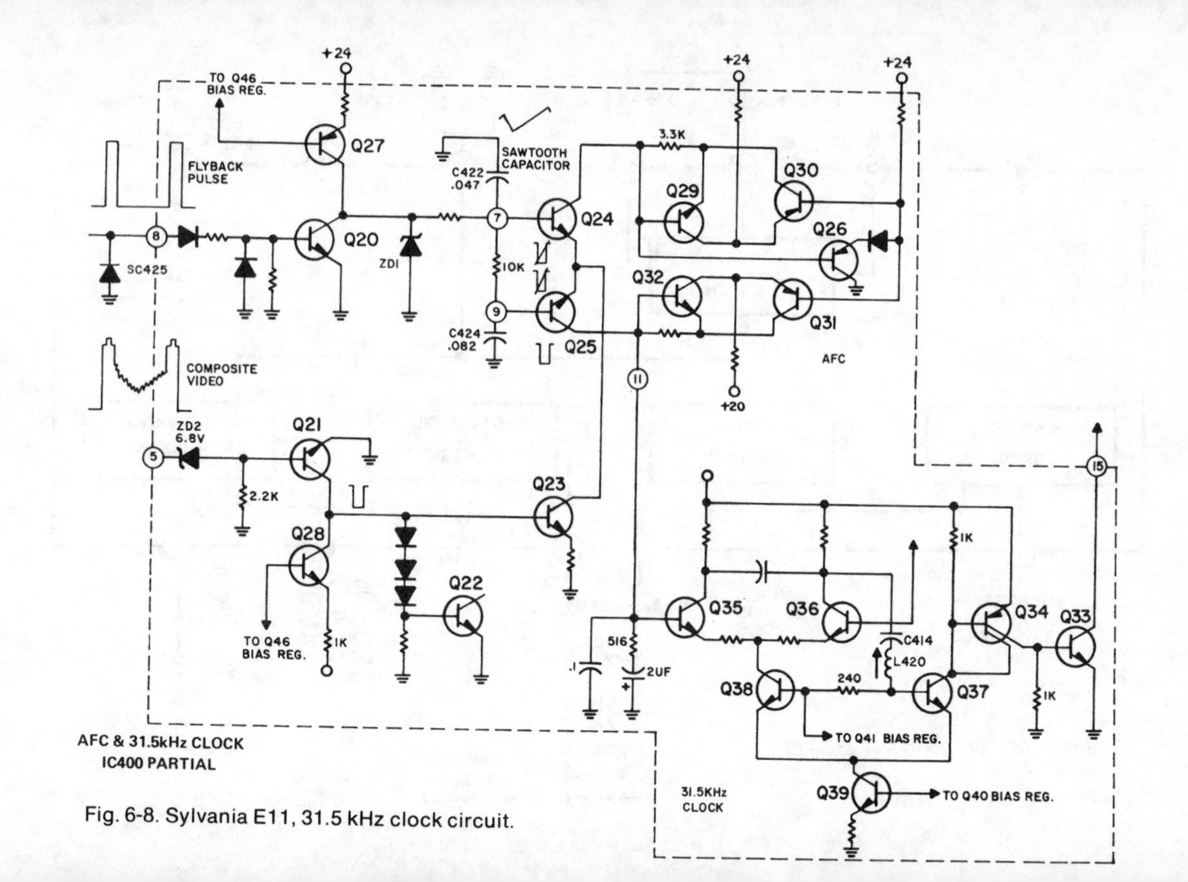

Fig. 6-8. Sylvania E11, 31.5 kHz clock circuit.

terminal 7. Capacitor C422 at pin 7 integrates the pulse into a sawtooth waveform, which is used by phase comparator Q24-Q25 for comparison with station sync pulses.

Sync pulses are separated from the positive sync, composite signal by ZD2, Q21, and Q28. The separated sync is amplified by Q23, then coupled to the emitters phase comparator Q24-25. When the sawtooth and the sync pulse are phase locked, no correcton voltage is developed, so the 31.5 kHz clock timing requires no change. However, when the sawtooth frequency is lower, the sync pulse rides down the sawtooth slope, reducing the differential amplifier forward bias, and speeds up the clock until the sawtooth and sync pulse are phase locked.

The clock (oscillator) frequency is 31.5 kHz, which is adjusted by L420 and C414, with AFC voltage applied to Q35 (part of the 31.5 Hz clock). Q37 feeds the oscillator signal to Q24, a dual-collector transistor that couples the clock pulse to pin 15 of IC400 through Q33.

Possible Cause: The first check in this circuit would be to test for any abnormal voltages at the pins of IC400. If all voltages check near normal, try replacing IC400 with a new chip. Should picture symptoms still persist, then use your scope to check all input and output signal pulses on the IC. There should be a composite video signal at pin 1 and a horizontal keying pulse at pin 8.

In this chassis, no keying pulse was found at pin 8. This, of course, caused a loss of horizontal sync and AGC action. Here a shorted SC425 diode caused the keying pulse to be lost. An open R428 resistor, coupling capacitor, or faulty winding on the sweep transformer would also cause a loss of this vital horizontal keying pulse.

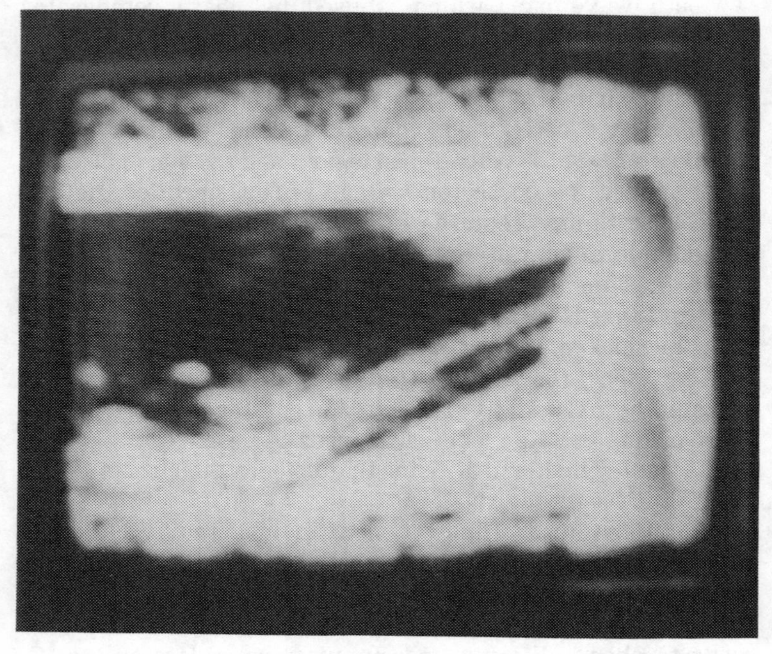

Picture Symptom: Picture rolls and will not hold horizontally. Look for trouble in the video processor module. May also look like an AGC overload problem.

Circuit Operation: An IC is used in the processor module (Fig. 6-9) to generate AGC, sync, and noise gating. The IC produces the AGC voltages as well as both positive- and negative-going sync information. Noise gating takes place as an internal function of the chip.

Composite signal information exits at pin 5, and it is at this point where video noise cancellation takes place. For any impulse noise higher than the sync amplitude, that portion of the composite video will be blanked out. The noise-cancelled video is fed back into pin 14 of the IC for sync separation through C401, C403, and diode CR401. These components form a dual-time-constant circuit that provides the sync limiting stages with immunity to plane flutter. After limiting, positive-going sync is available at pin 3 and negative sync at pin 2.

The internal AGC gating stage is turned on during the horizontal sync interval by a positive horizontal-sweep pulse

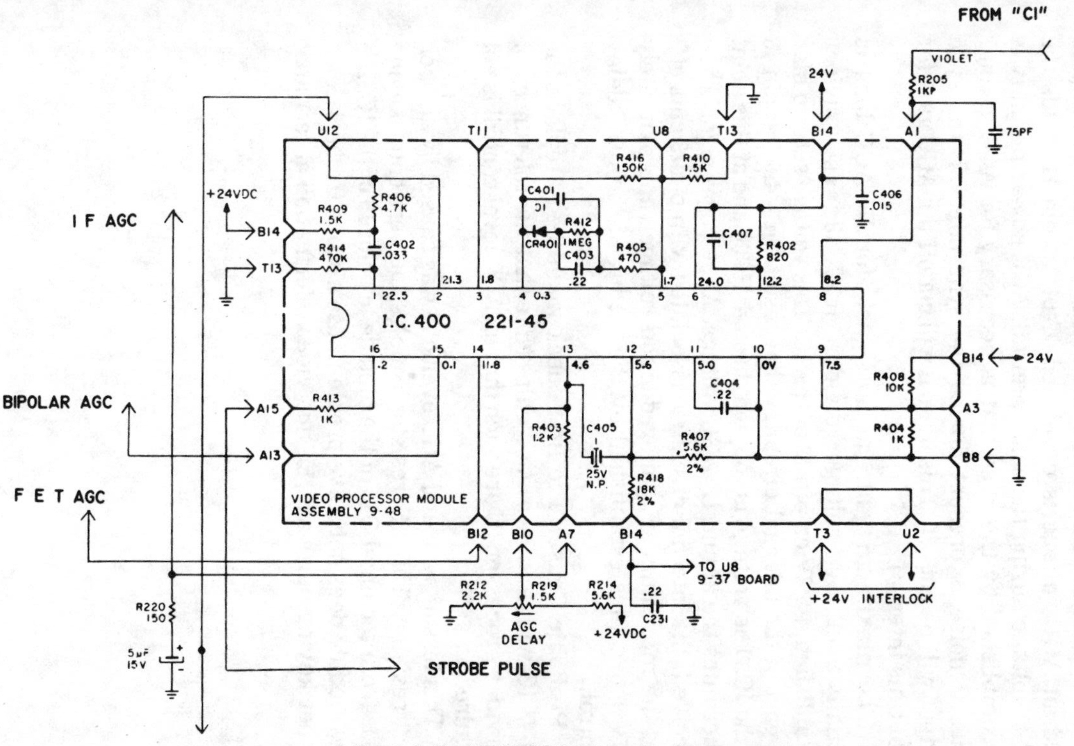

Fig. 6-9. Zenith 25CC25 chassis, video processor.

coupled to pin 16, and a portion of the negative sync output coupled through R406 and C402 to point 1. During gating time, the amplitude of the incoming horizontal sync pulse is sampled and an AGC voltage is developed. C404, at pin 11, holds the AGC voltage during the scan interval. AGC noise immunity is accomplished by the sync that is necessary for AGC gating; thus, if noise is present during the sync interval, the sync output will be cut off. This in turn will cut off the AGC gate for that time interval.

The maximum gain (at no signal) for IF AGC bias is initially set by voltage divider R407-418. The divider established +5.6V on pin 12 of the IC. This voltage sets up bias on part of the internal circuitry that results in +4.6V at pin 13 of the IC. The voltage is coupled through R403, and at this point it becomes the actual IF AGC voltage. As stronger signals are received, the internal circuitry associated with pin 13 cuts off, and the maximum IF gain-reduction voltage (approximately +7V) becomes established by the setting of the AGC delay control.

Probable Cause: The first check would be to replace the *211-45* IC with a known good chip. Note that the tolerances of the video processor IC are such that an AGC level control is not required.

This loss of sync and AGC overload was caused by an open R413 resistor. This 1K resistor couples the horizontal keying pulse from terminal Q15 of the module to point 16 of the IC. If a new module does not solve problem, use a scope to check for a proper keying pulse at Q15, and video information at terminal A1.

ZENITH 15Y6C15 CHASSIS

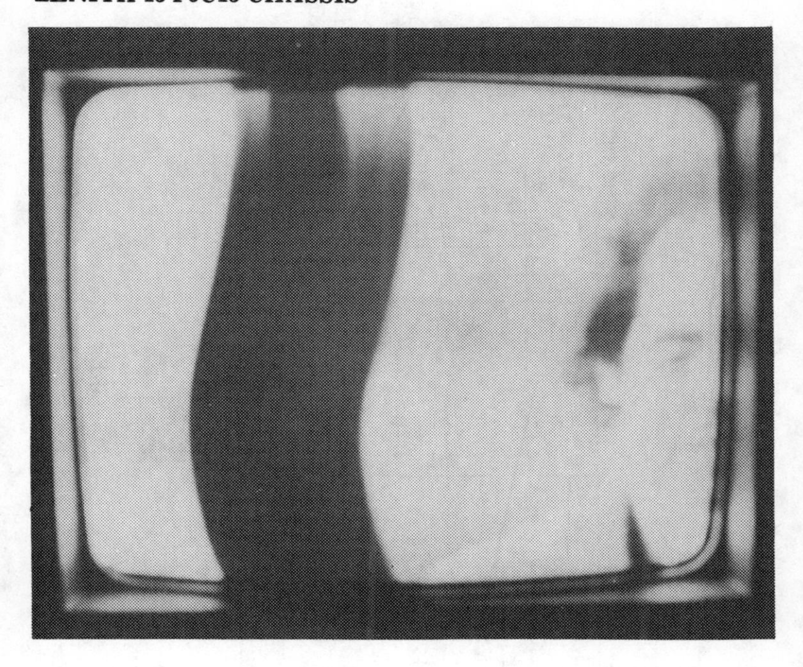

Picture Symptom: The picture has an AGC overload condition and a loss of sync. The AGC level and AGC delay controls had no effect on the picture. Look for trouble in the AGC system. The fault could also be located in the video IF, video detector stage, or AGC delay circuit.

Circuit Operation: This chassis has a transistorized IF system. A lower amplitude (about 4V peak) video signal is detected by the video detector. Tube-type IFs produce around 6V peak of video. Thus, the signal coupled to the grid (pin 7) of V4 for noise gating is also coupled from the plate of the cathode follower, and this helps boost the gain.

Sustained noise pulses or an input signal of high amplitude may cutoff V3A, which could cause the IF amplifiers to remain at full gain due to lack of AGC action. The result could be *lock-up*, which is a sustained overload condition that prevents proper AGC action. However, under such a high signal condition, the video coupled to sound-sync amplifier V3A will result in an increase in the screen voltage (pin 7), which is coupled through C67 as a positive "pulse," allowing V4 to

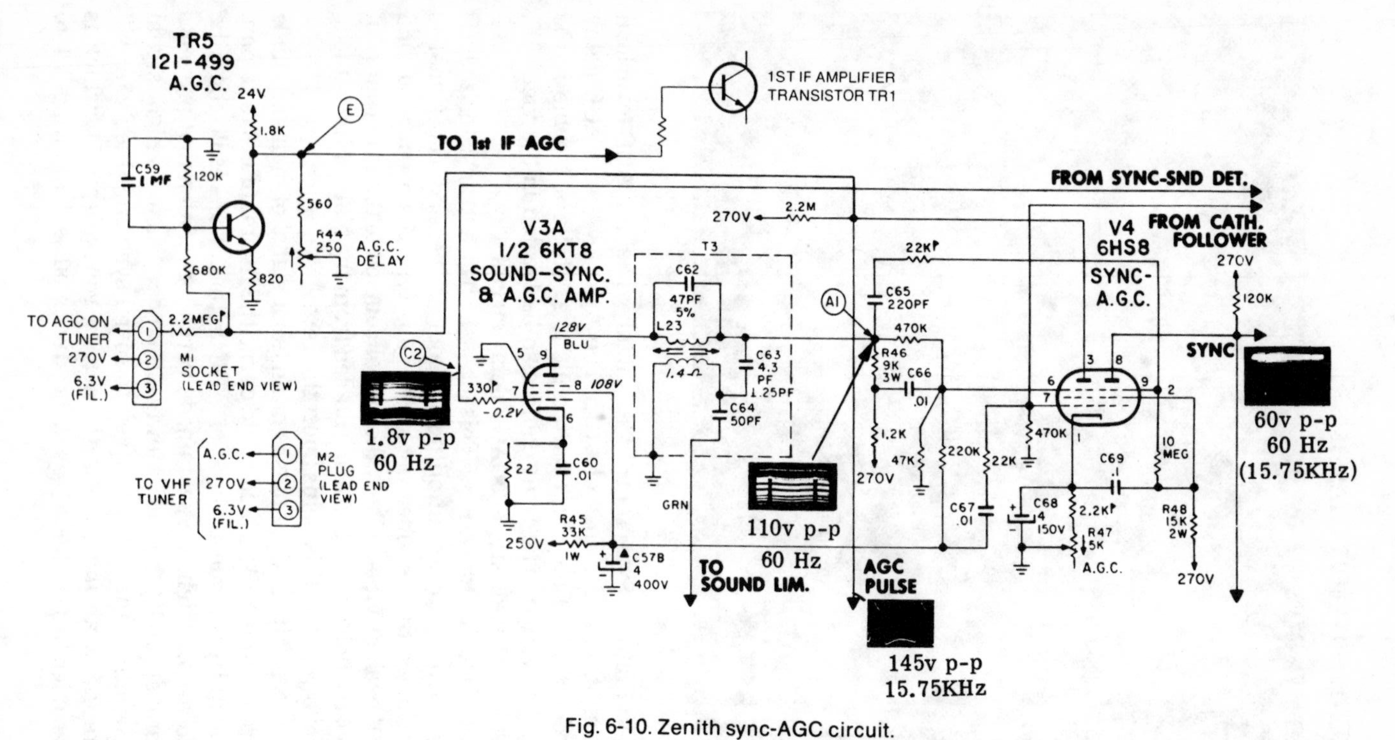

Fig. 6-10. Zenith sync-AGC circuit.

conduct heavily until AGC is again developed to decrease the signal. Thus, lock-up is prevented. Refer to the sync-AGC circuit in Fig. 6-10.

The negative (or positive) AGC voltage developed at the plate (pin 3) of V4 is fed to the tuner RF stage through a 2.2M resistor and to the base of TR5 through a 680K resistor. The AGC voltage for the first IF must be a positive voltage because of the NPN transistor. To decrease the gain of an NPN transistor amplifier, the base voltage must be increased. Thus, the AGC voltage produced by V4 must be used in a way that will control a positive AGC voltage for the first IF stage.

Under a given no-signal condition, AGC tube V4 is not conducting. Thus, the base of TR5 is 5 to 6 volts positive, as a result of the voltage divider consisting of a 2.2M resistor (from 270V at plate of V4), the 680K, and the 120K resistor to ground. The voltage at the tuner under this condition is about 1.0V positive. The 1 μF capacitor (C59) located in the base circuit of TR5 is a nonpolarized type, since the base voltage of TR5 may be negative or positive, depending upon signal strength.

Transistor TR5 is also across part of another voltage divider, consisting of a 1.8K, 560Ω, and 250Ω AGC delay control (R44). The positive base voltage of TR1 is about 4 or 5 volts, depending upon the setting the AGC delay control. An AGC delay voltage is required for transistor IF amplifiers.

With a weak signal condition, V4 will conduct and the voltage at base of TR5 will decrease to 2 or 3 volts positive (depending on signal strength). The tuner AGC voltage will be near +0.5V, or only very slightly positive. Transistor TR5 conducts less and the voltage at the base of the first IF will increase slightly, reducing its gain slightly.

With a strong signal, the tuner AGC voltage may vary and the base voltage of TR5 may now become negative, cutting off TR5 and producing a voltage of about +7.0V at test point E and the base of the first IF transistor stage TR1, reducing its gain. The range of AGC control in TR1 is adjustable by the setting of R44, the AGC delay control.

Probable Cause: This picture lock-up was caused by an open C67 capacitor. This is a 0.01 μF coupling capacitor that normally feeds a positive pulse to the control grid (pin 7) of V4.

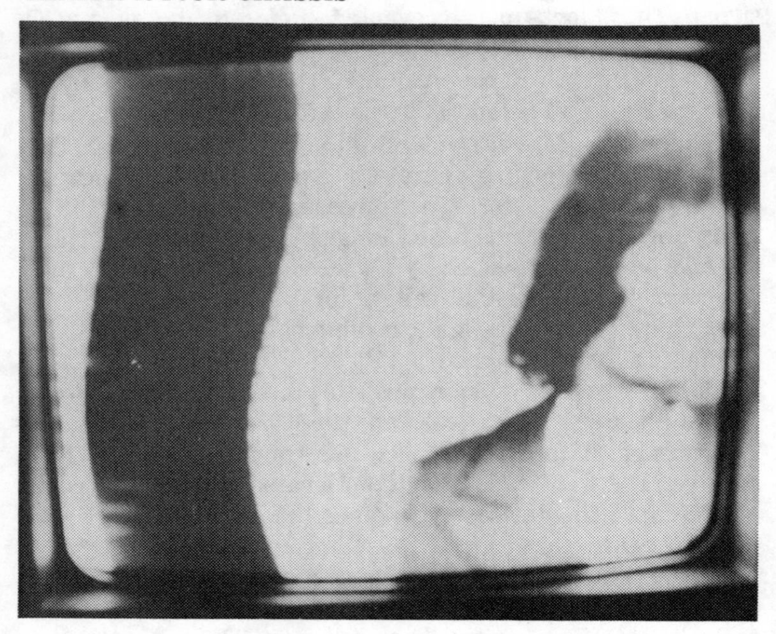

Picture Symptoms: No horizontal lock. The horizontal hold control could be adjusted to make the picture float across the screen, which indicated the horizontal oscillator could be set to correct frequency. Another symptom could be tearing or jittering as shown in Fig. 6-11. Look in the AFC circuit for loss of horizontal sync pulse, lack of comparison pulses from the horizontal sweep transformer or trouble in the horizontal phase detector circuit.

Circuit Operation: This is a popular horizontal-phase and oscillator-control circuit (Fig. 6-12) that uses a 6U10 triple-triode tube to generate and control the horizontal sweep frequency. There is also a pair of diodes used as a phase detector that accepts a pulse from the sweep output circuit and compares its phase (timing) with the incoming horizontal sync pulse.

From the sync separator comes a 70V P-P negative-going pulse (Fig. 6-13), which is fed into one-half of C140, used to filter out any vertical sync information. Should C140 become defective, this could cause a loss of sync pulses to the diodes or may let vertical sync pulses pass and cause unstable horizontal picture lock.

Low-amplitude horizontal pulses are found at the junction of diodes X8 (center of the two 470K matched resistors). The correct pulses are shown in Fig. 6-14 as taken at test point B. At this piont, sync pulses are compared in phase with a pulse taken from a winding of the horizontal sweep transformer. The correct feedback comparison pulse from the sweep transformer is shown in the bottom scope trace of Fig. 6-15. Any change in phase between the incoming sync pulse and feedback pulse from the sweep transformer will result in a DC correction voltage. The correcton output from the diodes is a modified sawtooth waveform that is shown in the top trace of Fig. 6-15, taken at test point C. This pulse is then filtered by a 1M resistor and capacitors C143 and C144. This is part of the anti-hunt network, and should C144 or C143 become open, the oscillator would keep changing frequency (hunting)—producing a piecrust appearance.

This DC correcton voltage is then fed to the control grid (pin 7) of the reactance control section (V10A). Any change of the DC control voltage causes the internal resistance of the reactance control tube to change also. This tube section then appears as a variable reactance across the horizontal

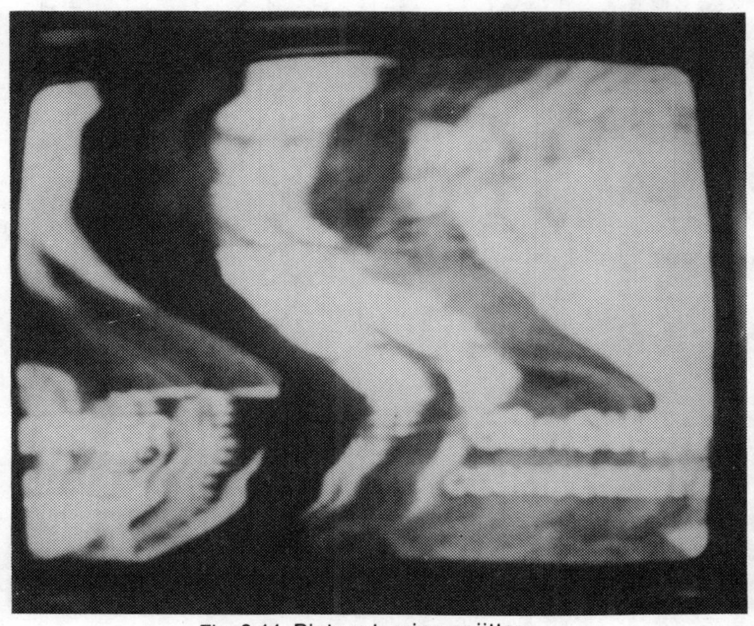

Fig. 6-11. Picture tearing or jitter.

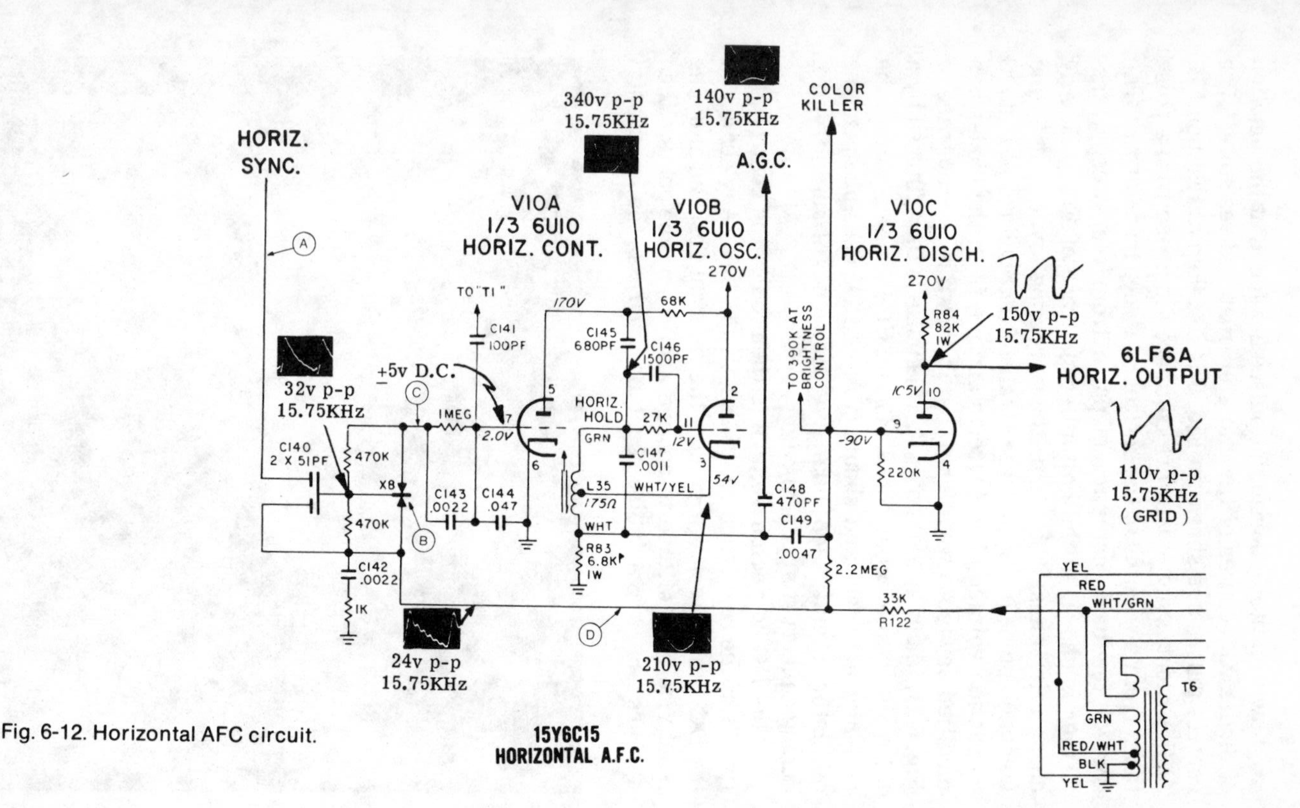

Fig. 6-12. Horizontal AFC circuit.

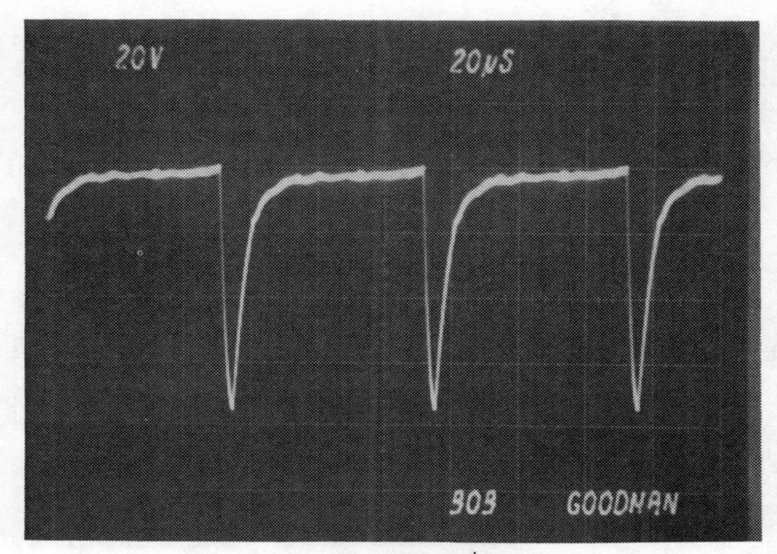

Fig. 6-13. Correct horizontal sync pulse.

oscillator tank, changing as a function of the DC correction bias or sync phase. This then controls the oscillator section (V10B), keeping it on frequency and in phase with the horizontal sync pulses.

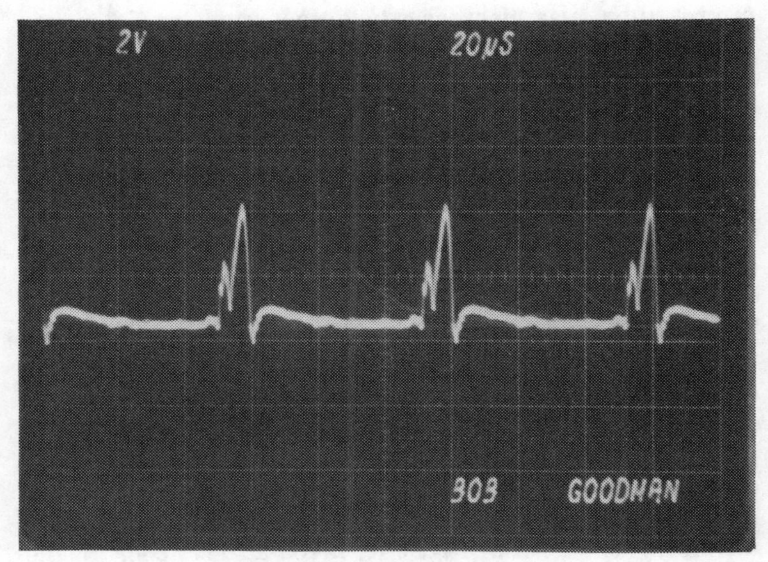

Fig. 6-14. Correct pulse at test point B.

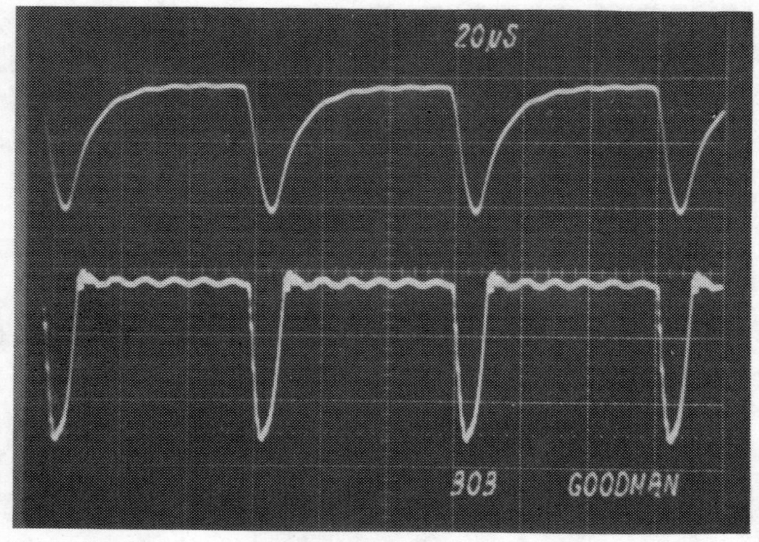

Fig. 6-15. Correct comparison pulse (bottom trace); modified sawtooth waveform (top trace).

When the DC output voltage from the phase detector goes more negative, the bias on the control tube is increased, resulting in a higher frequency output from the oscillator. As the DC voltage from the phase detector goes more positive, the control tube bias decreases and the oscillator frequency lowers. When the phase detector control voltage is at its normal operating point (in this circuit it happens to be −2V), and this is the case when the sync pulse and feedback pulse voltage are exactly in phase, then the control tube bias becomes constant and the horizontal oscillator frequency stabilizes at 15.75 kHz.

Probable Cause: The AFC dual-diode devices contain two closely matched diodes to assure electrical balance, but they will cause critical sync trouble should they become faulty. These AFC diodes are further matched with an external pair of resistors to balance the system. Make sure these resistors are also matched in value. The diodes should be checked by substitution for any type of horizontal AFC sync symptom. Forward and reverse resistance readings are not always conclusive—unless the diodes are obviously open or shorted. An open AFC diode will produce a high-amplitude pulse waveform as shown in Fig. 6-16, taken at test point B, the common cathodes of the diodes. If the phase detector diodes are replaced, make sure thay are not installed backwards. The picture may lock in, but the horizontal hole will be critical if the diodes have been reversed.

The loss of horizontal hold in this set was caused by a faulty (open) 33K feedback pulse resistor (R122).

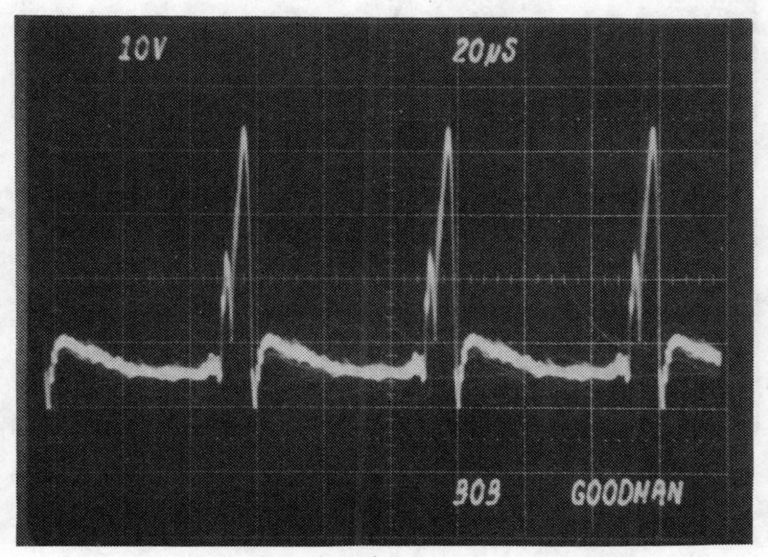

Fig. 6-16. Waveform when AFC diodes are open.

ZENITH 20Z1C37 COLOR CHASSIS

Picture Symptom: Picture blacks out when switched to a strong TV station. This is sometimes called an *AGC lock-up* condition. Look for trouble in the AGC system.

Circuit Operation: In this AGC circuit (Fig. 6-17) a VDR is used as a "sensitivity" device. An example of this VDR action can be seen by having a no-signal input to the receiver. Under this condition, the IF amplifiers are at maximum gain, so the AGC tube is essentially cut off. If suddenly a strong TV station is switched in while the IF amplifiers are at full gain, enough video amplitude is coupled to the AGC amplifier tube to cause it to cut off, thereby maintaining the high gain in the IF section and setting up a blocking or overload condition. But if the bias on the AGC tube could be changed momentarily to cause conduction, the AGC would again be developed and the IF and tuner RF amplifier systems would go back into operation. Thus, an automatic system is needed. (In some instances, poor B+ regulation will prevent this blocking condition, but will not always work.

In this circuit a VDR (voltage-dependent resistor) is used to produce an "automatic" bias change during the reception of a strong signal. When a strong signal cuts off the AGC tube, it

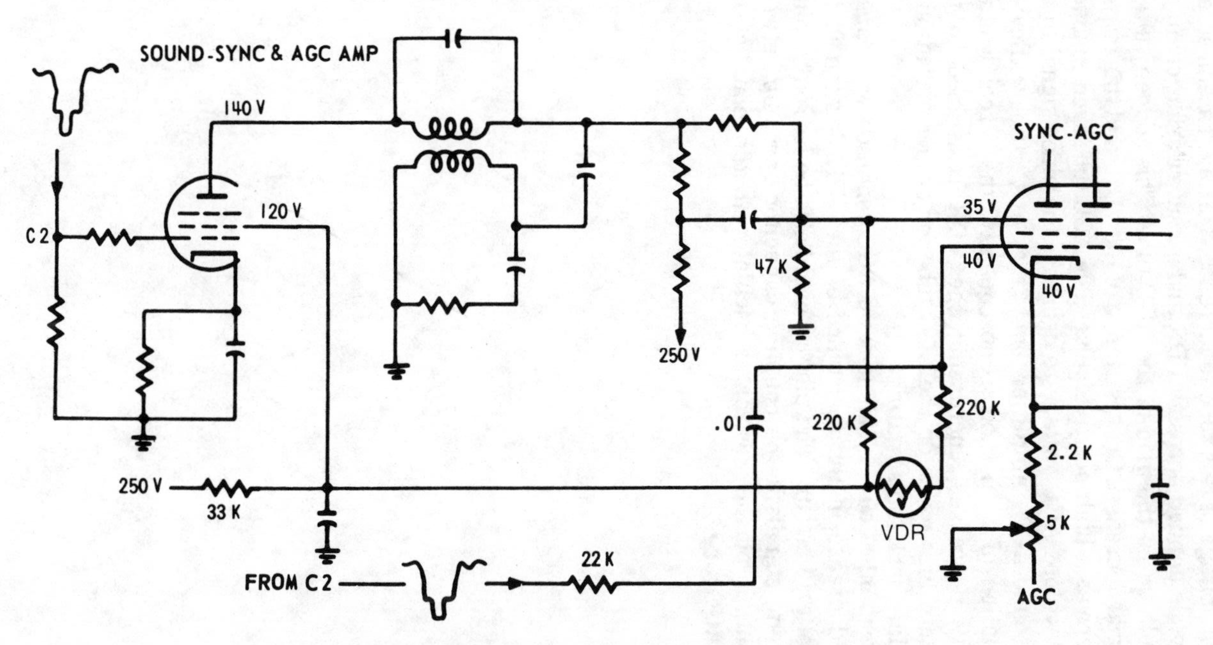

Fig. 6-17. Zenith 20Z1C37 chassis, sound and sync circuit.

also cuts off the sync-sound-AGC amplifier, which causes the screen voltage of this tube to increase. This increase in screen voltage is applied across a VDR and a 220K resistor, changing the voltage on the grid of the AGC tube. This sudden voltage increase decreases the resistance of the VDR, and this change appears as a high positive pulse. The AGC tube then conducts and develops AGC voltage to decrease the video signal to the sync-sound-AGC amplifier and the AGC tube, allowing conduction of both tubes. Consequently, the screen voltage rapidly decreases, increasing the VDR resistance to the optimum bias value on the AGC tube for proper operation of the fringe lock (noise gating).

Probable Cause: This lock-up AGC condition was caused by a decrease in the resistance of the VDR, located in the control grid circuit. The picture could also be lost on medium to strong signals if the coupling capacitor opens up between the output of the horizontal oscillator to the plate of the sync-AGC tube.

Picture Symptom: Picture is very weak or dim and has a weave. Look for trouble in the video processor module and its associated circuits.

Circuit Operation: This chassis utilizes a sophisticated sync and AGC system, which has been used in most Zenith chassis for several years. The *221-45* monolithic chip has a reference level that is internally generated and controlled by zeners, and this eliminates the need of an AGC level control. Refer to module circuit in Fig. 6-18.

Just like a conventional AGC circuit, this system must have a keying pulse and a negative-going separated sync pulse present in order to operate. Coincidence of the two pulses result in proper AGC action if the detected signal level changes. However, if the horizontal oscillator is out of sync, no AGC action will occur until the horizontal keying pulse coincides with the signal sync pulse. Thus, no sampling is accomplished during scan time. Also, since an AGC gating pulse is produced only by the coincidence of the sweep pulse and the separated sync, not only is the AGC comparator

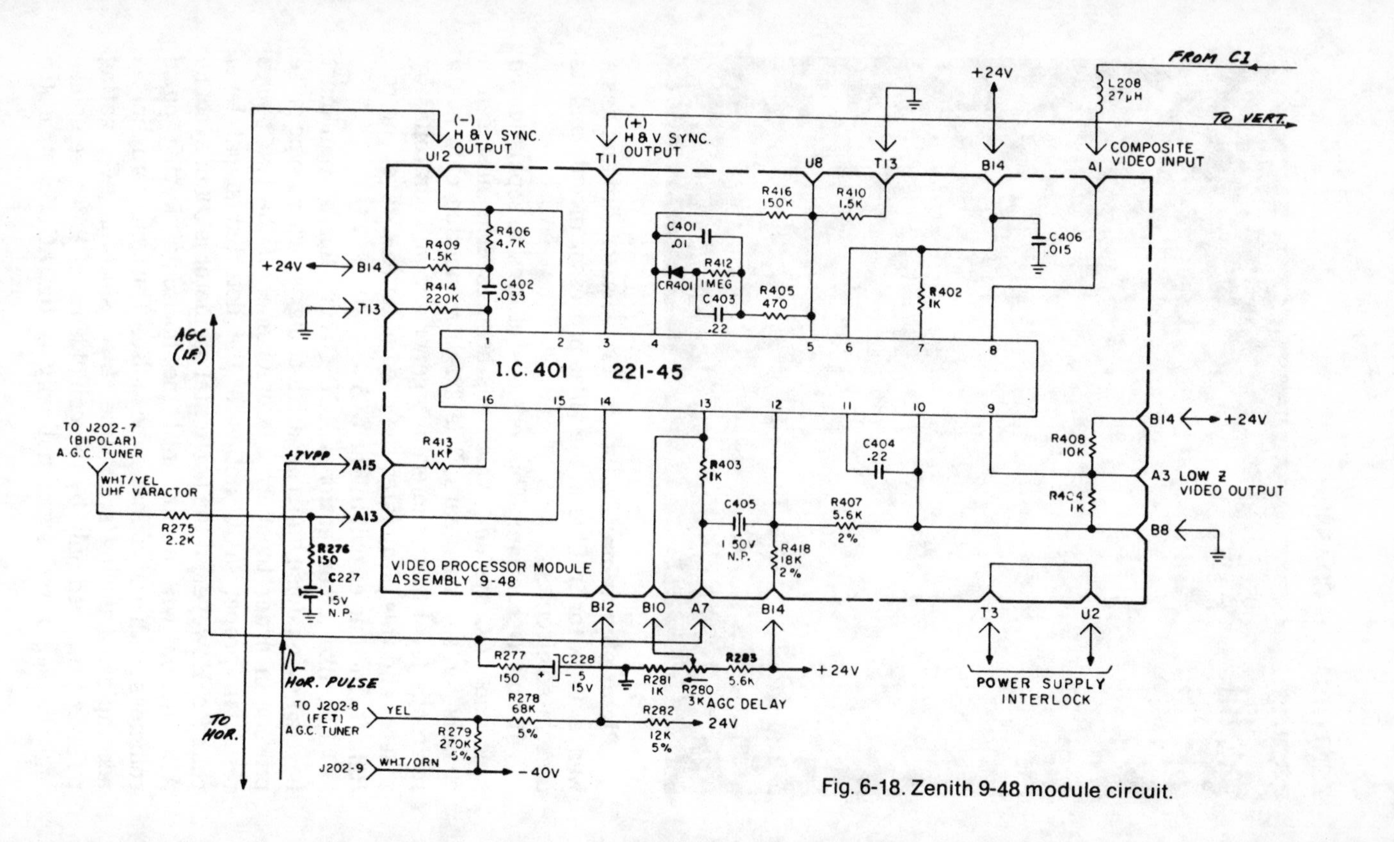

Fig. 6-18. Zenith 9-48 module circuit.

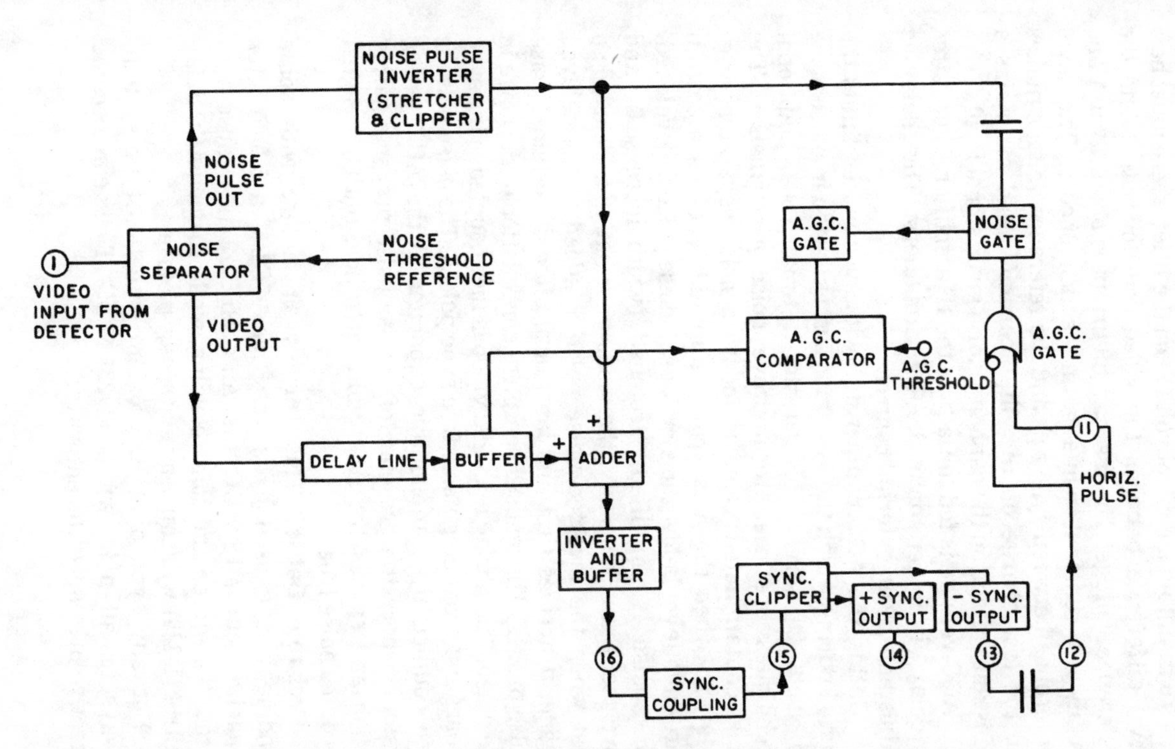

Fig. 6-19. Noise protection circuit.

activated but a constant current discharge of the AGC filter occurs.

To maintain a control voltage on the filter, there must be a level difference between the video detector output and the reference voltage. The discharge current is about 1.7 mA and is always constant while gating action occurs. The level difference between sync tips and the reference voltage makes up the lost charge on the filter capacitor and maintains a constant voltage on the capacitor for a given signal level.

AGC voltage is fed initially to the IF amplifier. Then, after a delay, it is fed onto the RF amplifiers. This delay is adjustable by an external control.

Gating the AGC comparator with sync pulses limits the period during which the system is susceptible to impulse noise to about 20 percent of the signal time. Thus, additional techniques are used to improve noise performance. The noise-contaminated video is fed to a noise separator (Fig. 6-19), so when the noise pulse amplitude exceeds a certain threshold level below the sync tips, a noise pulse is developed. This pulse is amplified, stretched (width increased), and inverted. Simultaneously, the video information is delayed so that when the inverted noise pulse is added to the delayed signal, it arrives first and lasts until after the original noise pulse passes. This causes complete cancellation of the noise in the video. The noise-cancelled video is then coupled to the sync separator, producing noise-free sync pulses. The noise pulse is also coupled by a monolithic capacitor to the AGC gating system, where the presence of a noise pulse prevents gating action and causes the AGC voltage to hold until the signal becomes noise free.

Probable Cause: This very weak picture with some bending was caused by a very leaky C228, a 5 μF filter capacitor on the IF AGC line. An open C228 may cause bars across the screen and some bending, while an open AGC delay control (R280) will cause a very snowy picture.

For any sync of AGC problem with this type system, always install a known good *221-45* IC. They can become defective in many different ways.

ZENITH 4B25C19 CHASSIS

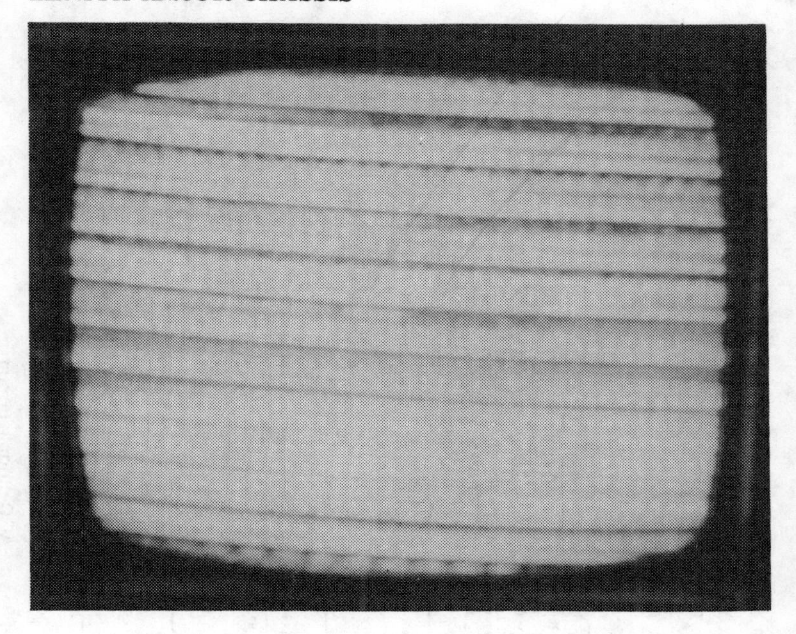

Picture Symptom: Loss of horizontal and vertical sync. Look for trouble in the *9-23* sync-AGC module.

Circuit Operation: The *9-23* module shown in Fig. 6-20 is used for generation of sync and AGC voltage. The "clipped sync" is used for synchronization of the horizontal and vertical oscillators, while AGC voltage is developed to provide a nearly constant detector output voltage. Because this is a closed-loop system, it becomes more advantageous to service and observe the sync-clipping action first.

Sync clipping is the action of removing from the composite video those components not required for the synchronizaton process. These unwanted portions are as follows:

1—All video information.
2—All blanking information.
3—The top and bottom sections of the actual horizontal and vertical sync pulses.

The top section of the pulse is undesirable because noise pulses would be present at this point, while the bottom part would be too close to the blanking and video areas. Only a small portion contains the sync pulse.

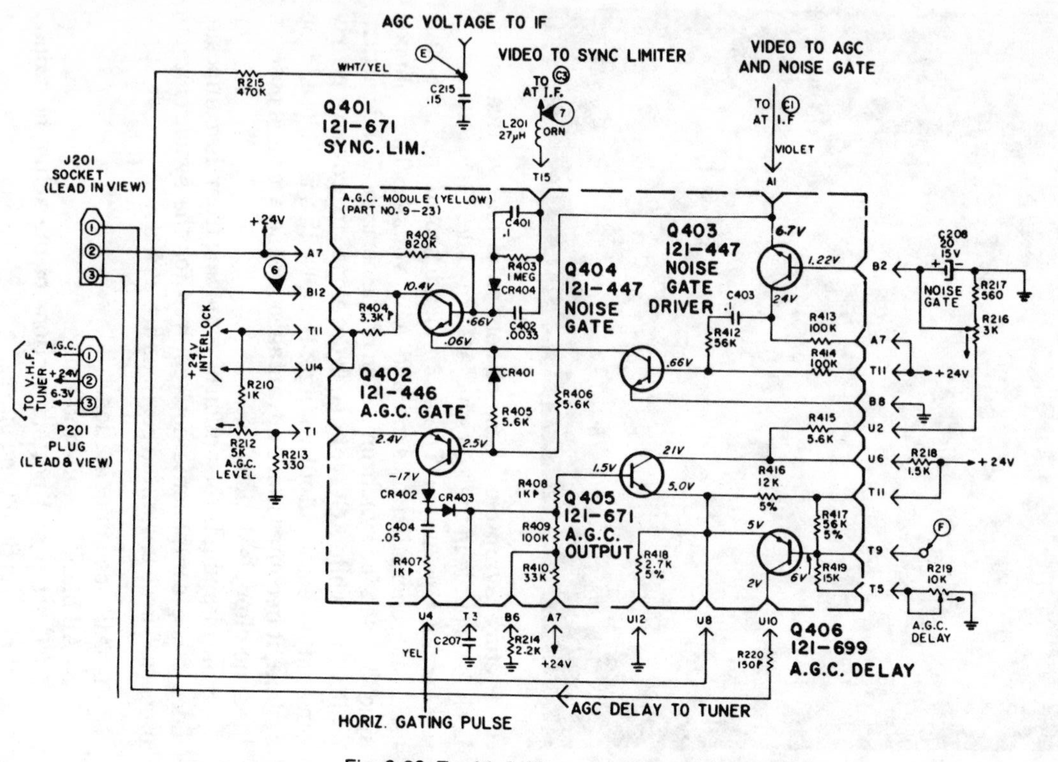

Fig. 6-20. Zenith 9-23 sync-AGC module circuit.

In most Zenith TV chassis the sync-clipping action is coupled to the AGC action. The pulse in Fig. 6-21 shows why this is so. If the DC signal level changes, or the amplitude of the signal gets larger or smaller, the predetermined clipping action is defeated. Sometimes it will include the noise pulses on top of the sync; other times it may bite into the blanking or even the video portion. These actions should be considered when troubleshooting both sync and AGC circuits.

AGC troubles can be detected by observing the sync information at the output of the sync clipper. The only active device in the sync-clipping action is Q401, an NPN transistor. Input signals are fed through a dual-time—constant filter to the base. With proper biasing and transistor parameters, clipping will occur and the clipped sync will be observed on terminal B12, the collector of Q401.

To make things more interesting, noise immunity circuits have been added around the sync clipper—Q403 is the noise driver and Q404 is the noise gate. (Note the redrawn circuit in Fig. 6-22.) These transistors and their associated circuitry have the specific function of making the horizontal and vertical oscillators less susceptible to noise pulses. However, since these circuits are connected to the clipper, they can, if defective, cause a sync-clipper problem.

Probable Cause: This loss of sync was due to a fault in the noise immunity circuits. Noise gate transistor Q404 was found to be open, and when this happens it actually removes the ground return from the emitter of Q401, the sync limiter stage.

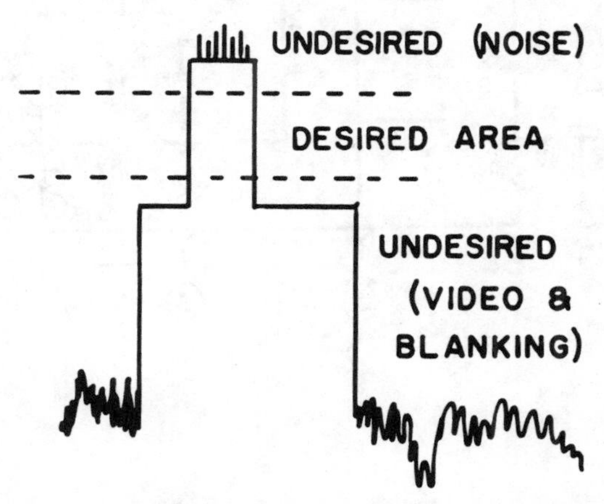

Fig. 6-21. Drawing showing the desired area of the sync pulse.

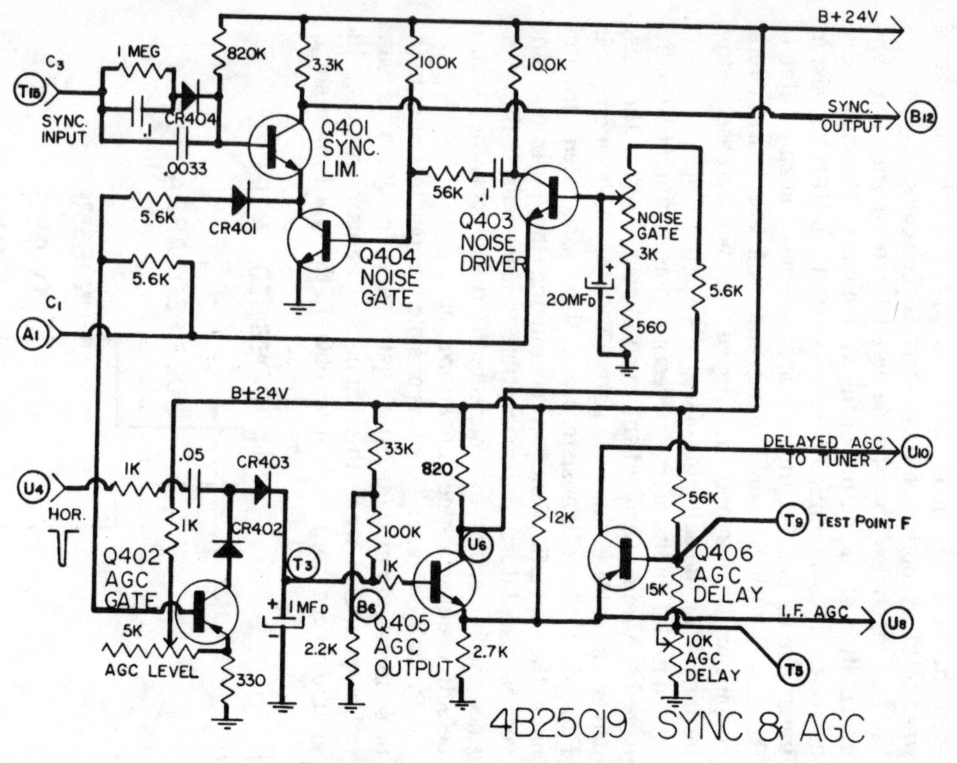

Fig. 6-22. Sync-AGC circuitry redrawn.

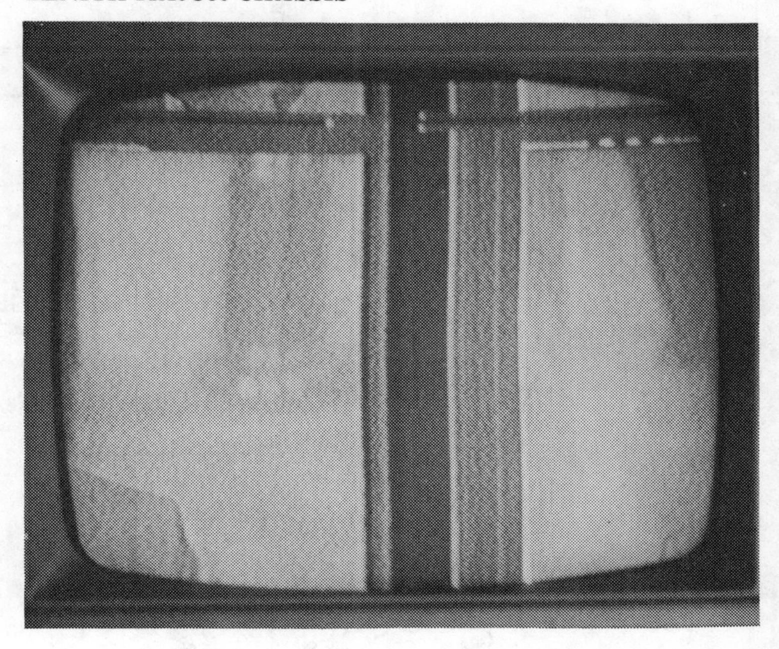

Picture Symptom: Picture floats around the screen which points to loss of vertical and horizontal sync. Loss of vertical and horizontal sync indicates trouble in a *common* sync circuit, so let's check the progress of the sync signal in the common stages.

Circuit Operation: This chassis uses a 6BA11 tube that is a combination sync clipper, noise canceller, and AGC control (Fig. 6-23).

A negative-going composite video signal is coupled to point 4 of V204A. This usually has a low amplitude, and with no noise on the signal, it has little effect on tube conduction. When a noise pulse does appear, a negative-going spike sends the tube deeper into cutoff for the duration of the noise burst and prevents distortion of the sync pulses at pin 6.

A composite video comes from the sync and AGC stage (V203A) and is fed to the pin 7 grid of V204A to remove the vertical and horizontal sync pulses. The average conduction of the sync separator is controlled mainly by the screen grid (pin 3). The operating point is set so that only the upper 30 percent of the video signal can bring the tube out of cutoff, permitting

Fig. 6-23. Zenith 14A9C50 chassis, sync-AGC circuit.

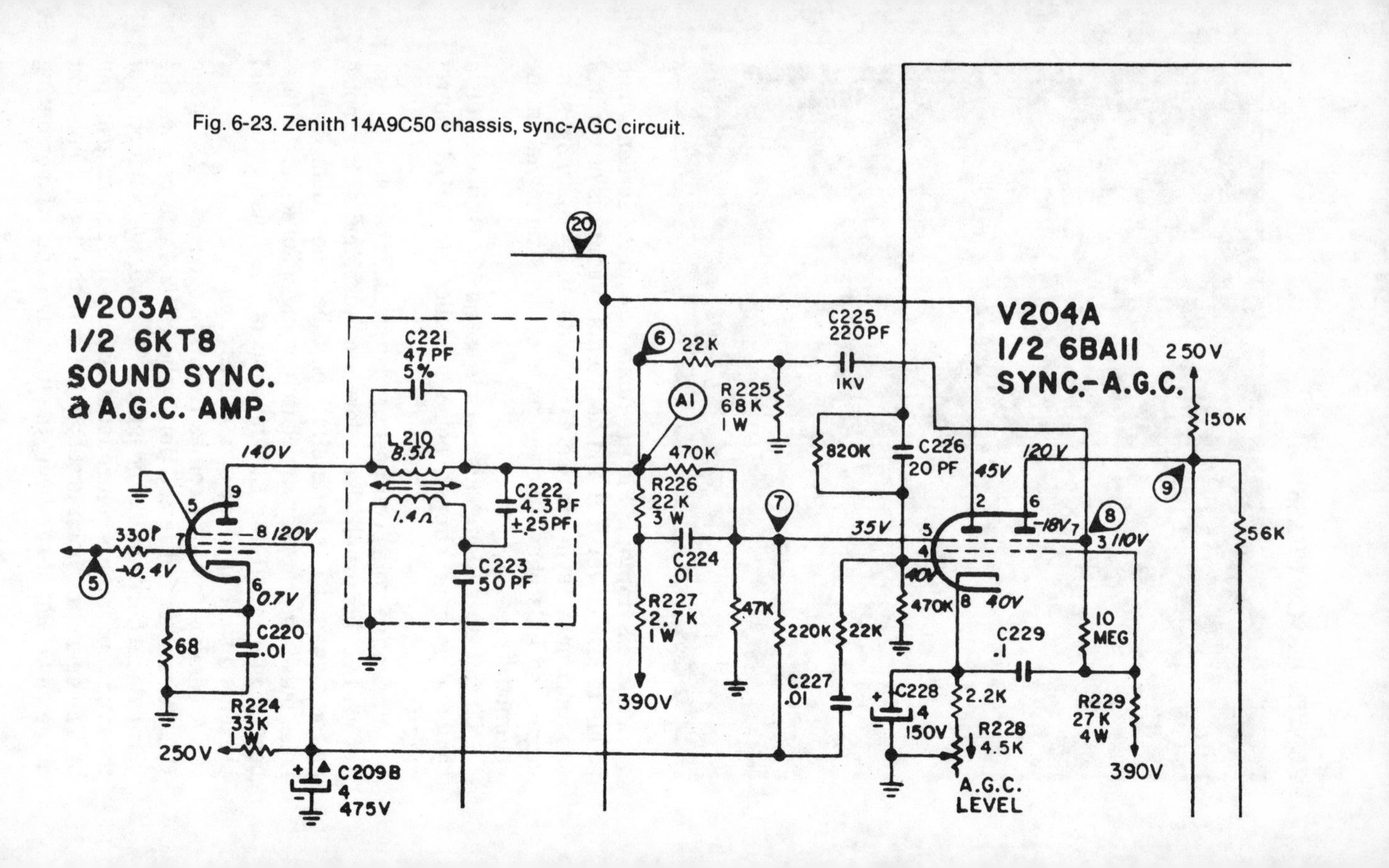

only sync pulses to appear in the output. As the AGC control in the cathode circuit is set for minimum resistance, the grid bias between pins 4 and 8 goes more positive and the tube conducts more. Thus, the AGC voltage at pin 2 (plate) goes more negative and causes IF cutoff.

Circuit Checks and Service Tips: Let's look at a few ways to check out this three-way circuit. The trick is to find out which section is defective, since one fault usually masks the other operations. The first grid (pin 4) is part of the noise cancelling circuit. A good way to start the check is to short together pins 4 and 8 with a clip lead. If the set will now operate, the trouble is in the noise canceller circuit. Use a scope to check out the circuit to find out why it is not working. Check to see if the DC bias voltages are correct. For the AGC section, use a scope to check for proper horizontal-keying and sync pulses.

Probable Cause: For sync troubles in the case sited, check for correct DC voltages at plate pin 6 and control grid pin 7. Then scope the same points for composite video and sync pulses. A leaky C225 (220 pF) coupling capacitor caused the grid voltage at pin 7 to measure +5V, thus upsetting bias on the clipper and cutting off sync pulses at the plate. This caused a complete loss of vertical and horizontal sync.

ZENITH 15Y6C15 CHASSIS

Picture Symptom: No horizontal picture lock. The picture could be made to float back and forth, which indicated that the horizontal oscillator could be set on frequency. Look for loss of horizontal sync pulse, lack of comparison feedback pulses from the horizontal sweep tranformer, or trouble in the AFC horizontal phase detector circuit.

Circuit Operation: This popular horizontal sweep circuit (Fig. 6-24) uses a 6U10 triple-triode tube to generate and control the horizontal sweep frequency. A pair of diodes is used as a phase detector that accepts a pulse from the sweep output circuit and compares its phase (timing) with the incoming horizontal sync pulse. The horizontal sync is a 70V P-P negative-going pulse fed into one-half of C140, which is used to filter out vertical sync information. A faulty C140 could cause a loss of sync or let vertical sync pulses pass, causing unstable horizontal picture lock. Low-amplitude horizontal pulses will be found at the junction of diodes X8. It is at test point B that sync pulses are compared in phase with the horizontal feedback pulses. Any change in phase between the incoming sync and feedback pulse will result in a modified

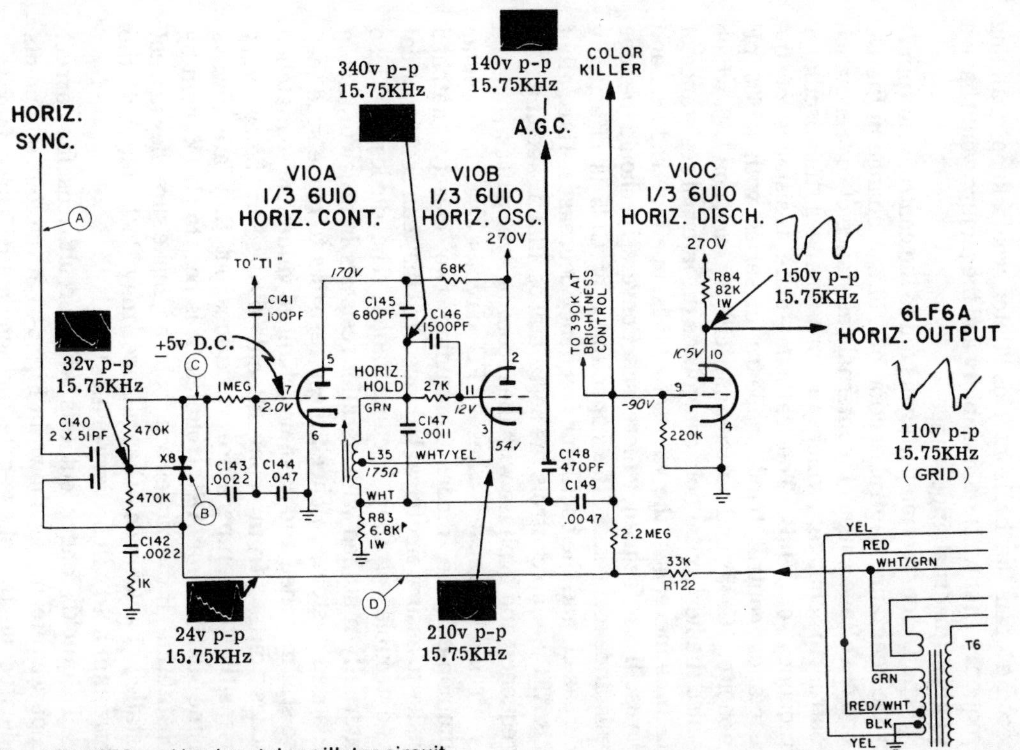

Fig. 6-24. AFC and horizontal oscillator circuit.

sawtooth waveform at test point C. This pulse is then filtered by a one megohm resistor and capacitors C143 and C144 to produce the DC correction voltage. These components are also referred to as the *anti-hunt* network. Should C144 or C143 become open the horizontal oscillator would keep changing frequency (hunting) quite rapidly and the picture would have a wiggly or piecrust appearance.

The DC correction voltage is fed to the control grid (pin 7) of the reactance control section of V10A. A change in the DC control voltage causes the internal resistance of the reactance control to change, which in turn changes the oscillator frequency of V10B. The cathode–plate resistance of the reactance control triode appears in series with a 680 pF capacitor (C145) and horizontal oscillator tank coil L35, which is also the horizontal hole control. As the internal resistance of the tube increases, the value of C145 is, in effect, reduced and the oscillator frequency increases. Conversely, should the tube resistance decrease, the capacitance of C145 is effectively increased, and the oscillator frequency decreases. Thus, all of this AFC action controls the horizontal oscillator, keeping it on frequency and in phase with the incoming sync pulses.

Probable Cause: These AFC circuits use the popular common-cathode diodes containing two closely matched diodes to assure electrical balance. And they can cause critical sync troubles if they become faulty. The diodes are also externally matched with parallel resistors in order to have a balanced system, so check these resistors to make sure they are still matched and have not changed value. These matched diodes differ from conventional diodes in being very insensitive to temperature variations but they are always prime suspects for AFC troubles. It is best to check them by substitution. If the diodes are replaced, make sure they are not installed backwards—the picture may lock in, but the horizontal hold will be very critical.

To quickly check out this AFC circuit, look for correct scope waveforms as shown in Fig. 6-25 at the four test points indicated on the phase-detector schematic. These four test points will give a clue as to which AFC actions are not operating properly.

To check horizontal oscillator operation for correct frequency, clip a jumper lead from test point C to chassis ground to defeat the AFC correction voltage. Now see if the

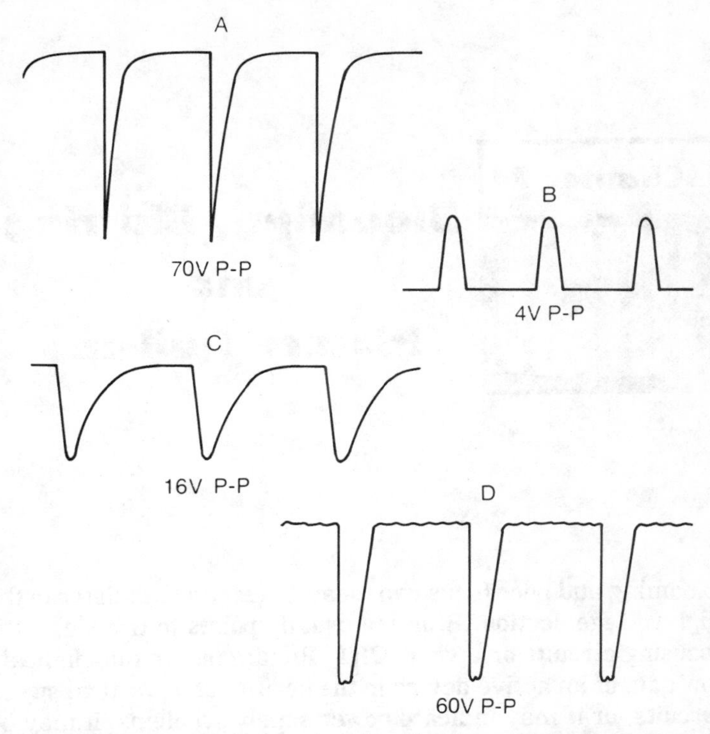

Fig. 6-25. Correct scope pulses that should be found at test points A, B, C, and D in Fig. 6-24.

horizontal oscillator can be adjusted to have a full picture that will float across the screen. If the picture can be made to slowly float, this indicates that the horizontal oscillator can be adjusted to the correct frequency, so the fault will be found in front of test point C, in the AFC or sync stages. If the picture cannot be made to float across the screen (only diagonal lines can be seen), this indicates an off-frequency horizontal oscillator rate, so the oscillator circuits must be checked out.

Also note that poor color sync may result from a fault in the horizontal AFC circuit, since this may change the phase of the color burst-gating pulse (timing) in some color chassis.

The loss of horizontal hold in this set was caused by an open R150 resistor (47K) feeding the horizontal comparison pulse to the AFC diodes.

Blooming, Blurring, and Picture Pull-In

Blooming and poor focus can mean a defective regulator in the high-voltage section. Blurring usually points to trouble in the focusing circuits or a weak CRT. Picture pull-in may indicate low gain of an active device in the horizontal or vertical sweep circuits, or it may indicate power supply problems. It may be helpful to question the set owner as to other possible symptoms.

Check the brightness and contrast controls when these problems seem related to the complaint. Do they affect the picture? Is any one color predominant in the picture? Does the symptom come and go, or is it continuous? Each answer should bring you closer to the solution, thus saving you time and keeping more bucks rolling in.

Most of the problems described in these case histories are significant, as they tend to show what corollary problems develop with specific principal complaints. You may be surprised at the uniformity of problems that can be traced to specific symptoms.

Picture Symptom: A very bright picture with foldover, or a bright narrow picture. Look for trouble in the horizontal sweep and B+ boost circuits.

Circuit Operation: When output transistor Q206 (Fig. 7-1) is turned on by a pulse from the pulse generator, it allows current to flow from chassis ground, through Q206 and the primary of T204 sweep transformer, to the +135V line. This current flow induces a voltage in the secondary of T204 and causes current to flow in the yoke, increasing until Q206 is turned off. During this time, the CRT electron beam moves from the center to the right-hand side of the screen. This is one-half of the trace portion of the sweep cycle.

When Q206 is turned off by the pulse generator, the energy stored in the magnetic field of the yoke causes the yoke to act as an energy source. The primary of T204 now acts as a secondary, and the voltage across the primary charges C234, and also charges C236 through boost rectifier Y208. The voltage polarities are such that damper Y206 is reverse biased. During this time, the electron beam moves from the right edge of the screen, back to the center.

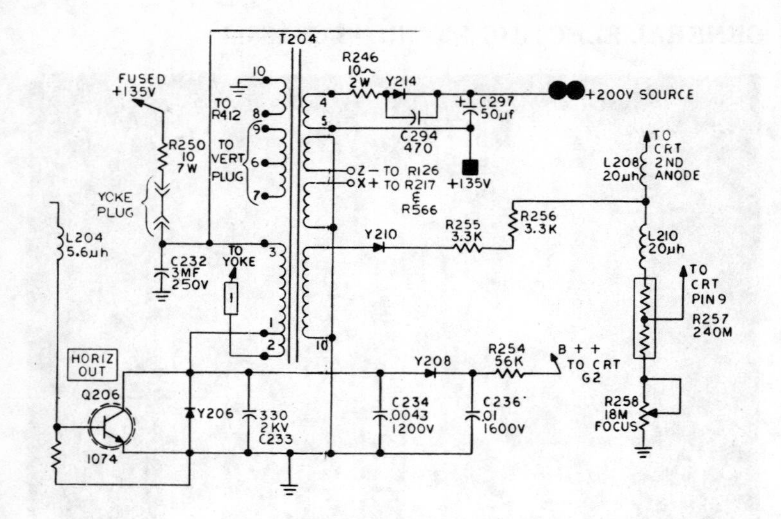

Fig. 7-1. GE QB chassis, horizontal sweep circuit.

Probable Cause: A very bright picture with possible foldover is caused by an open C234 in the boost circuit. An open C236 causes a very dim picture.

Picture Symptom: Very narrow picture in center of screen as shown in (Fig. 7-3) Sound is good and video looks all right. There is also a ripple effect caused by the horizontal sweep problem that could be deflection yoke trouble, but in this case it was not. Look for a fault in the horizontal sweep and centering circuits.

Circuit Operation: A simplified horizontal deflection and control circuit is shown in Fig. 7-2. The horizontal centering system consists of diode Y1101, resistors R1111, R1112, R1113, and coil L1101. This network is connected to points 1 and 2 of a winding on the horizontal sweep transformer and between the low end of the horizontal yoke coils and terminal 9 of the sweep transformer primary winding.

The horizontal pulse voltage between pins 1 and 2 of the sweep transformer (HVT) is scan rectified by diode Y1101 and develops a small voltage that if measured from point 9 of HVT will be negative at the R1111 end of R1113 and positive at the R1112 end. Thus, by means of the horizontal centering control R1113, the voltage between the low end of the deflection yoke and pin 9 can be varied from a little positive to a little negative. The voltage difference between these two points

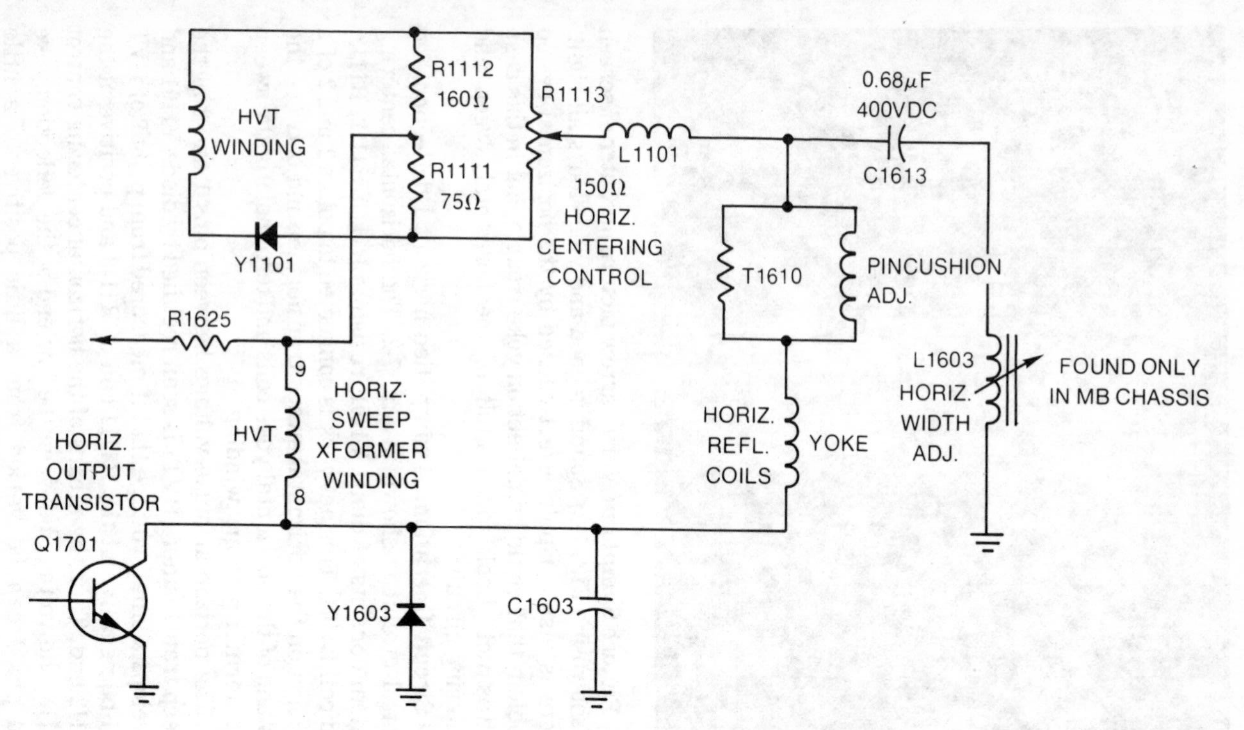

Fig. 7-2. GE MA/MB horizontal deflection system.

causes a current to flow through the yoke coils, and the amount and direction of that current is what moves the raster about. Coil L1101 presents a high impedance to AC through the deflection yoke while presenting a low impedance to DC. In the MA chassis the AC return path for the yoke is through C1613. Note that in the MB chassis this AC return path is through C1613 and horizontal width coil L1603 to ground.

Probable Cause: The very narrow raster shown in the picture symptom was caused by an open C1613 capacitor, an 0.68 μF unit with 400-volt rating. This capacitor was found shorted with no picture and a tripped circuit breaker. In the MB chassis, only an open horizontal width coil L1603 would cause the same narrow picture symptom.

The following components, if open, will also cause a loss of horizontal deflection just a thin narrow vertical line down center of screen: deflection yoke, pincushion transformer T1610, coil L1101, horizontal centering control R1113, and both R1112 and R1111 open at the same time. An open or shorted Y1101 diode would shift the picture over to one side.

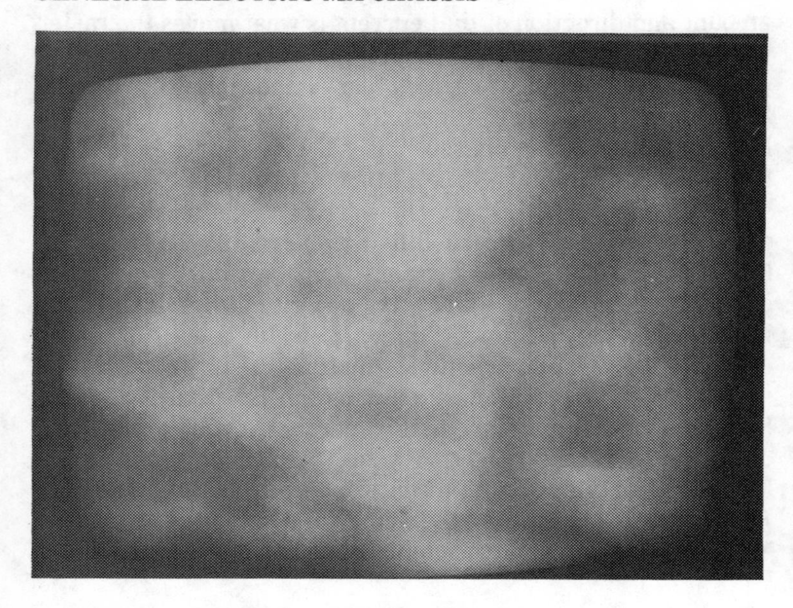

Picture Symptom: Video was very dim and looked out of focus. Predominant splotches of red and blue color were also noticeable. Looked somewhat like a purity problem, but was not. Look for trouble in the low-level video module.

Circuit Operation: Delay line driver Q310 in Fig. 7-3 isolates the video stages from the IF module and acts as an impedance-matching device to the input of delay line DL301. The output of DL301 is terminated by R305. The output of emitter follower Q302 drives the base of first video amplifier Q303 through the contrast control. Diodes Y301 and Y302 clamp the video to a DC level determined by the amplitude of the composite sync.

The signal is amplified again by Q304, the reference shift amplifier, in which negative horizontal and vertical pulses are introduced as blanking pulses to cut off the picture tube during retrace time. These pulses then clamp the video independently from the sync in the composite video, which feeds through the contrast control. The video signal is then fed through isolation emitter follower Q305 and capacitor C314 to the RGB amplifier module.

Zener Y346 is employed as a threshold element of beam limiter control R340 and Q306. The CRT beam limiter, Q307, is

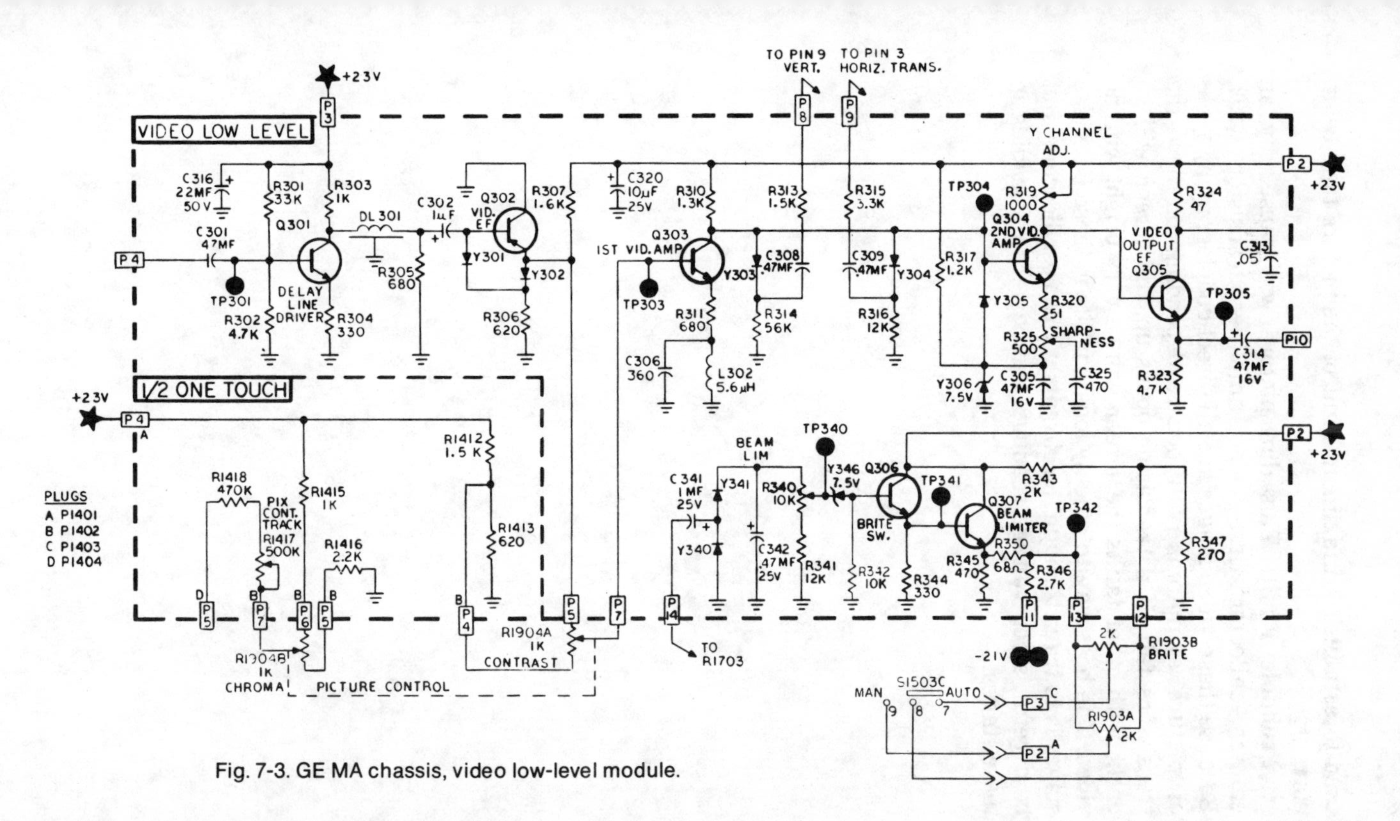

Fig. 7-3. GE MA chassis, video low-level module.

used to establish a maximum limit of the picture tube beam current.

Probable Fault: This dim picture was caused by an emitter–collector first video amplifier transistor Q303. Because these video stages are all direct coupled, many types of picture symptoms will occur, depending on how the transistors fail. If a transistor opens up, the screen will go dark; with other faults the screen will be very bright with no video. The best way to quickly locate faults in these DC video stages is to trace through the circuits with an oscilloscope. But be careful, since voltage readings can be very misleading in these stages.

MOTOROLA TS914 CHASSIS (QUASAR)

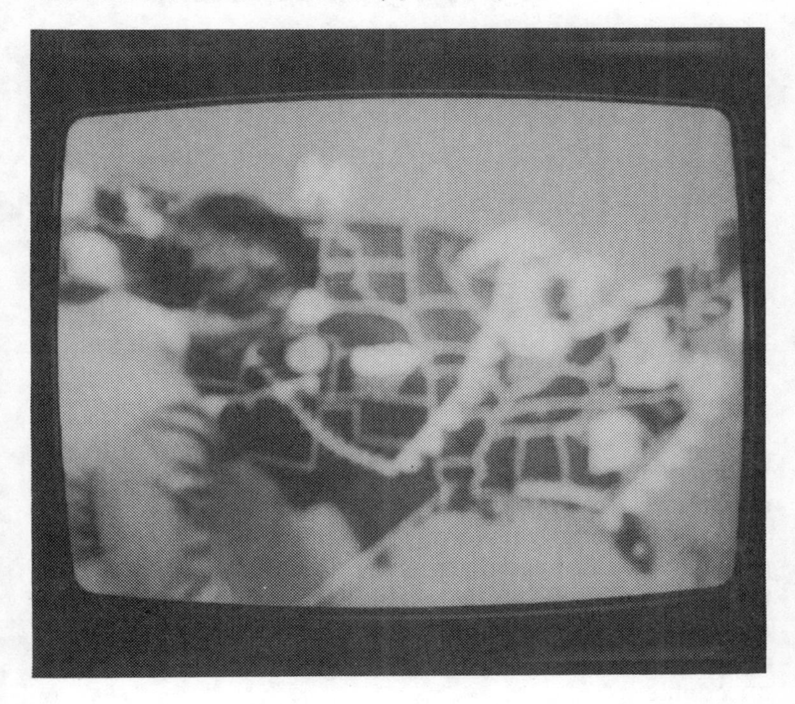

Picture Symptom: Purity was very bad and the picture was way out of focus. Could be defective picture tube or trouble in the focus control circuit. In this case the focus voltage was measured at 8 kV. This is way too high and could be caused by an open manual degaussing switch.

Circuit Operation: In this chassis (Fig. 7-4) the focus control goes to ground through R1 and the degaussing coil, except during degaussing. For the focus circuit this is the same as grounding the focus control, so correct focus voltage is obtained. The actual current through the coil is quite small and has no degaussing effect.

When the CRT needs degaussing, the switch is moved to the open positon for a moment. At this time the focus voltage goes up, because the focus control now returns to ground through the 22M resistor and C1. The increased voltage across R1 charges C1 2μF and when the switch is closed, the charge on C1 is fed back to the degaussing coil. C2 and the degaussing coil thus form an LC tuned circuit, and the shock of the DC

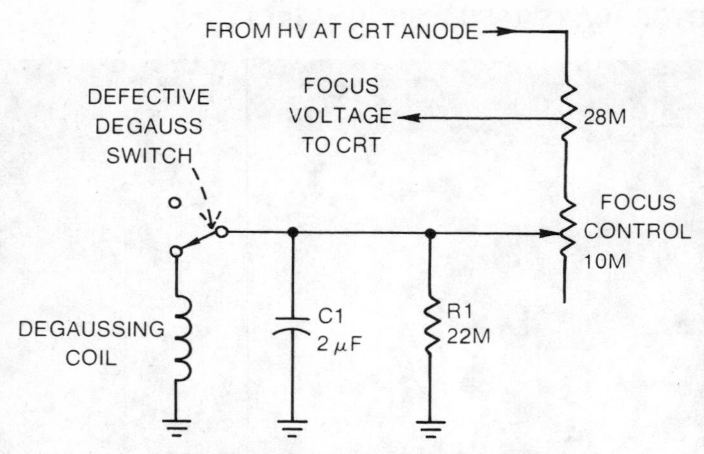

Fig. 7-4. Quasar TS914 degaussing circuit.

voltage being fed into the coil starts the circuit into a damped oscillation.

This damped waveform is a series of sine waves that gradually return to zero and in the meantime have degaussed the picture tube. When the switch is closed, the focus voltage decreases to normal, until the next degaussing time. So the open switch increases the focus voltage and makes degaussing impossible.

Probable Cause: Defective (open) manual degaussing switch. Replace the switch and adjust the set for good focus and proper purity.

Picture Symptom: At times the picture would bloom and be slightly out of focus. At other times the picture would be normal. A high-voltage probe was used to check the CRT anode and this voltage was found to vary from 22 to 31 kV. It looked as if this chassis had an intermittent problem in the HV regulator circuit system.

Circuit Operation: This RCA chassis uses a shunt HV regulator circuit (Fig. 7-5). The circuit is appropriately named because the regulator tube is shunted across the HV supply for the CRT. The shunt regulator maintains a constant high voltage to the CRT by keeping a constant load on the HV rectifier tube. The plate of regulator V102 is connected to the CRT HV anode, and the cathode to +405 volts. The control grid of the regulator is tied in with the HV adjustment control that receives voltage from the 850V B+ boost supply.

The B+ boost voltage varies with the amount of current being drawn from the HV supply and being fed to the picture tube. When the picture tube draws more beam current, the B+ boost voltage goes down and the regulator grid becomes more

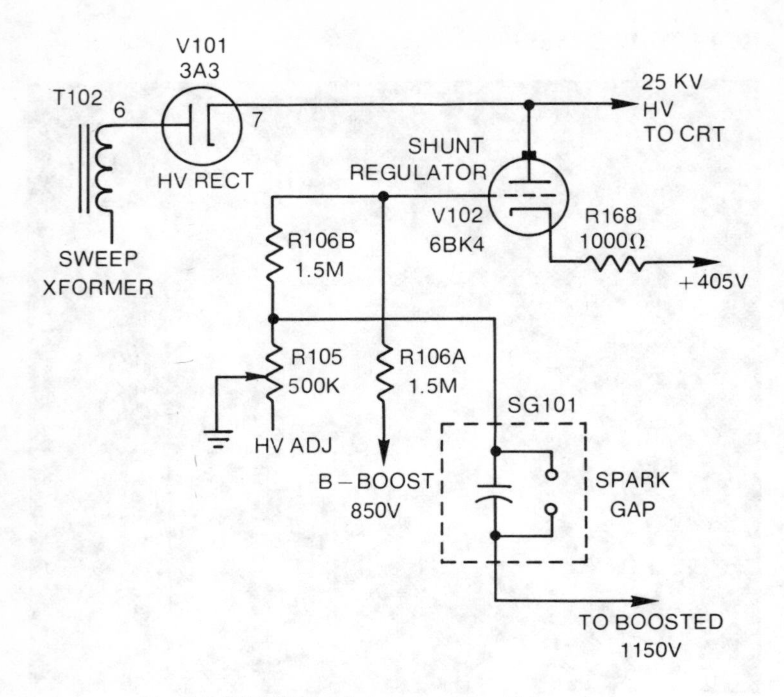

Fig. 7-5. RCA CTC-17 chassis, shunt HV regulator.

negative, which reduces the regulator current. With less CRT current drawn, the B+ boost goes up, causing the regulator current to increase. The total current (that is, current drawn by the CRT and the shunt regulator) stays about the same value during variations of the CRT loading, and therefore the voltage also stays about the same.

In normal operation the 6BK4 regulator tube draws no current when the CRT is adjusted for maximum brightness, and the regulator tube draws maximum current when the CRT is cut off. This then maintains a constant load on the HV supply and results in a constant HV output, regardless of the CRT beam current and brightness.

The simplified drawing in Fig. 7-6 shows that HV pulses from the sweep transformer are rectified by the HV rectifier tube. The regulator tube is connected in parallel (shunt) across the high-voltage output, and regulates by presenting a constant load to the HV supply. The CRT is also connected across the HV supply, and presents a variable load, depending on the picture image and where the brightness control is set.

As the brightness control is advanced, the CRT draws more current, and if there were no regulator, a large voltage drop would occur across the HV supply circuit. However, with a regulator system, the voltage remains constant, regardless of how much CRT current is drawn (up to a certain point).

Probable Cause: A defective regulator tube can give several different trouble symptoms. A tube with an open filament or no emission will not regulate, and the high voltage will climb up to around 30 kV or more. Note that in some older model RCA chassis (CTC-12, etc) a faulty regulator tube can cause vertical roll. A gassy tube, which will usually give off a purplish glow causes erratic HV regulation, lowers the HV more than normal, and probably will burn out the HV rectifier tube. A shorted shunt regulator will kill the HV, then cause the horizontal output tube to glow red, and if not fused will burn out the horizontal output tube along with the sweep transformer.

Referring back to the regulator schematic, some circuit faults could be as follows: R168, the 1000Ω cathode resistor, may increase in value and cause very little regulation action. A faulty HV adjust control, R105, will usually show up as control over the HV level. If R106A goes up in value the HV will be too high, and if R106A decreases in value the HV would be

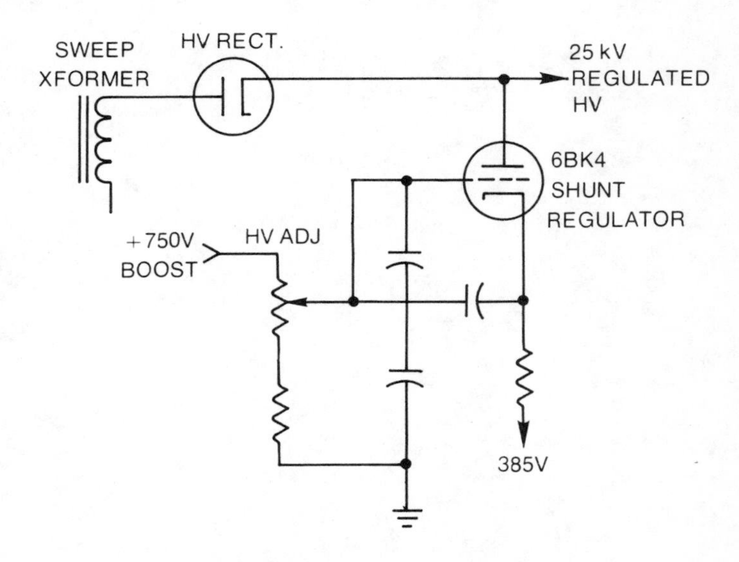

Fig. 7-6. Simplified shunt regulator circuit.

too low. Some chassis have a capacitor placed between the cathode and control grid of the regulator tube; if this capacitor shorts you will find no regulation taking place, and should it become leaky the high voltage level will go up and down in an erratic manner.

The picture symptom in this RCA CTC-17 chassis was caused by a faulty spark gap SG101, which intermittently put some of the 1150V boost voltage onto the control grid of shunt regulator V102, causing the HV variation.

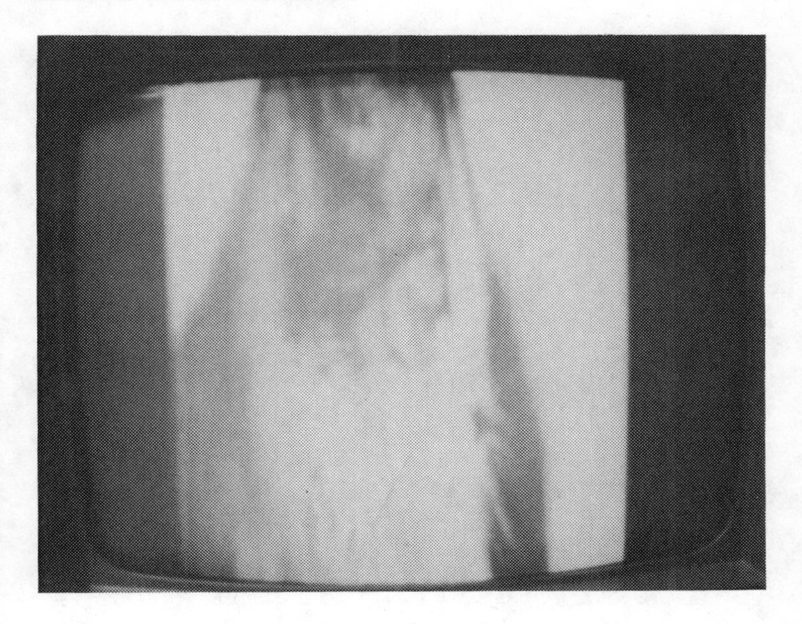

Picture Symptom: Picture pulled in from both sides. The high voltage at CRT anode was low, checking out at around 18 kV. Look for horizontal drive or HV regulation circuit troubles.

Circuit Operation: The HV regulator circuit (Fig. 7-7) has a positive pulse applied via capacitor C264 to the VDR (R333). The 1500V P-P pulse is obtained from a section of the primary winding of the sweep transformer.

A VDR (voltage-dependent resistor) is uniquely used in this circuit as a "rectifier," since a VDR (being a nonlinear device) will perform as a rectifier to any nono-sinusoidal waveform. The resistance of the VDR is voltage dependent, and in this case will offer a lower resistance to a higher voltage across it, and a higher resistance to a lower voltage across it. Thus, when the peak of the positive pulse is fed to the VDR through C264, its resistance lowers and causes the capacitor to charge rapidly to a relative high value. During the bottom portion of pulse voltage, the VDR resistance is higher and the capacitor tends to maintain its charge during this portion (trace time). The resulting negative potential appearing

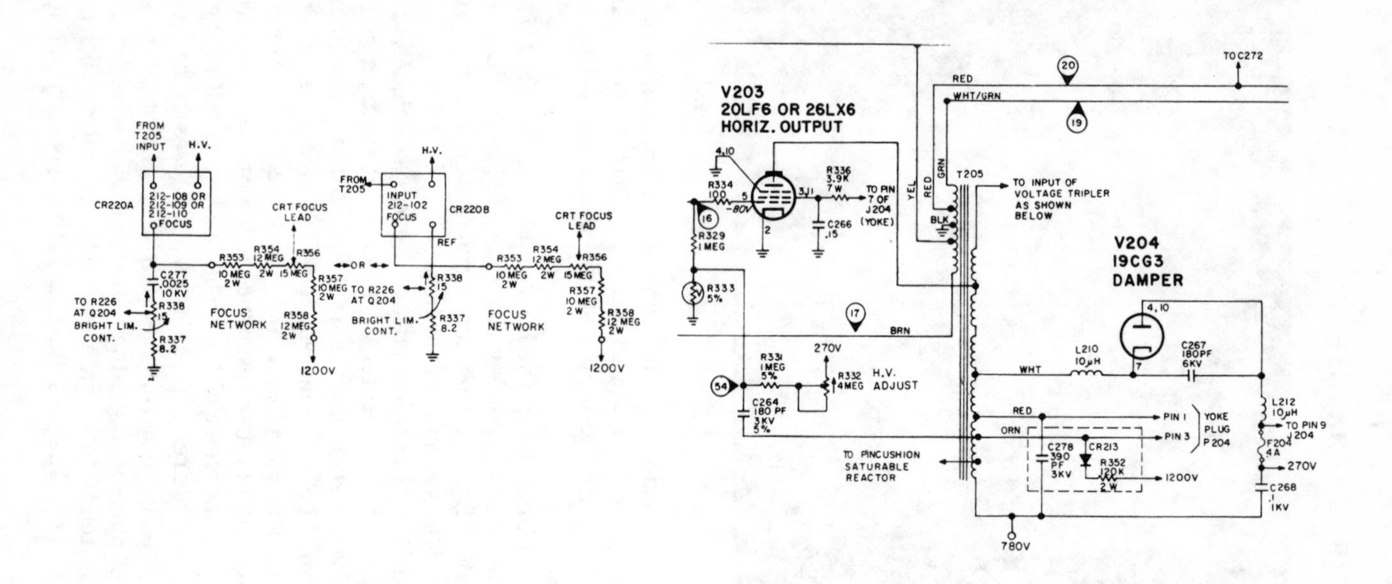

Fig. 7-7. Zenith 4B25C19 chassis, HV regulator circuit.

across R333 is applied to the horizontal ouput tube grid through resistors R329 and R334. Thus, the negative voltage adds to the negative potential at the grid.

If the HV increases, the positive pulse to capacitor C264 increases. In turn, the resulting negative voltage across R333 increases and the grid voltage of the horizontal output tube becomes more negative, lowering stage gain and thus lowering the output pulse to the sweep transformer.

To provide a means of adjusting the HV to correct value, the 270V B+ is coupled to the VDR through HV adjust control R332 and resistor R331. This control can be adjusted for a specific amount of positive voltage to counteract the negative potential from capacitor C264 and provides a means of setting the drive bias on the horizontal output tube for a specific value.

Probable Cause; This narrow picture was caused by an open HV adjust control (R332) located in the holddown circuit. A faulty VDR (R333) may cause the same picture problem.

ZENITH 15Y6C15 CHASSIS

Picture Symptom: The picture looked about normal, but would bloom when the brightness control was adjusted. A check of the HV at the CRT anode revealed that the chassis lacked any HV regulation. Adjustment of HV regulator control R112 had no effect on the high voltage.

Circuit Operation: A DC feedback voltage from the sweep transformer is used to control the control grid bias of the horizontal output tube and thus regulate the high voltage. Looking at the circuit in Fig. 7-8, note that a pulse of about 1500V P-P is fed from a primary winding of the sweep transformer via capacitor C153. This pulse causes the capacitor to charge from ground through the VDR (voltage-dependent resistor) to a specific value and is paralleled by an 8.2-megohm resistor and HV adjust control R85. After the pulse voltage diminishes, the capacitor discharges through the same components. But with a decrease in voltage across it, the VDR resistance increases and causes a higher resistance discharge path. So capacitor C153 tends to maintain its charge during trace time. This voltage is applied

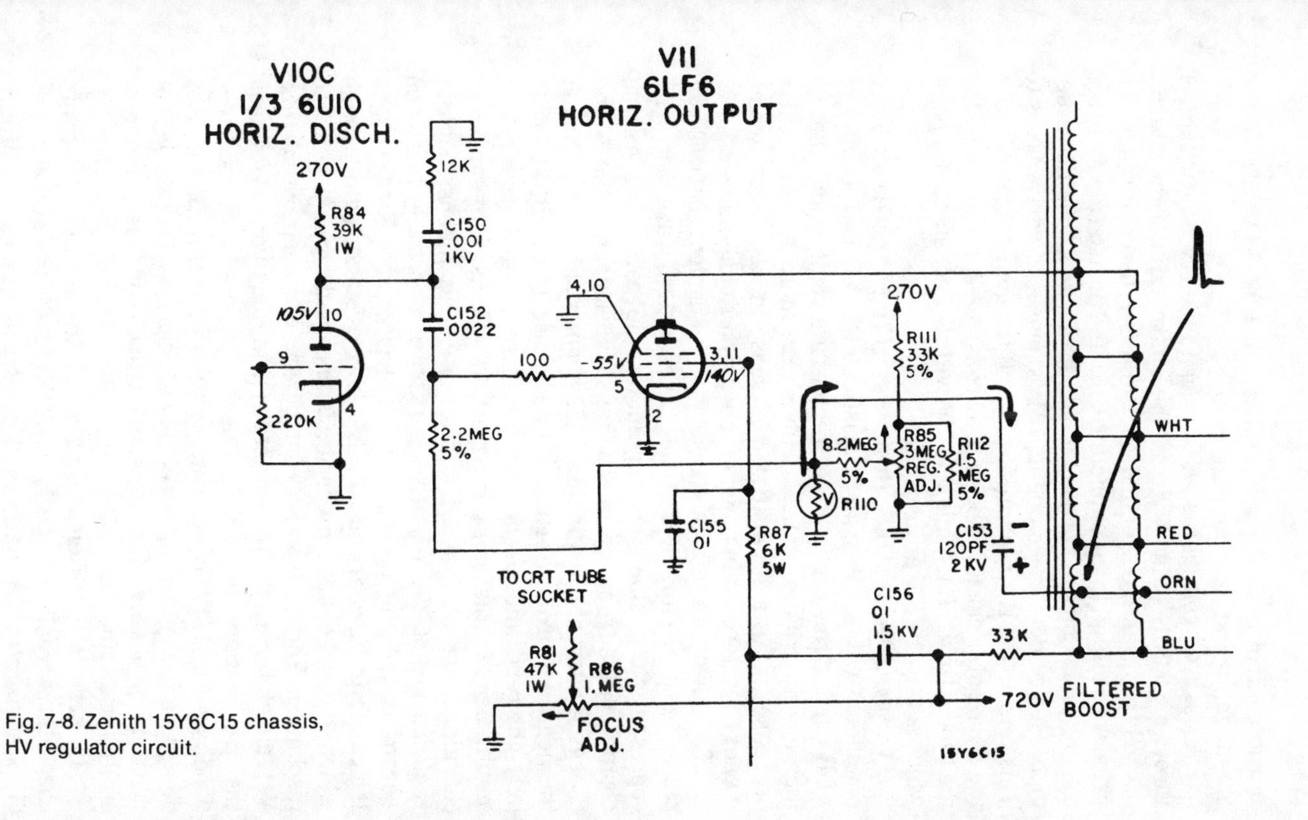

Fig. 7-8. Zenith 15Y6C15 chassis,
HV regulator circuit.

to the control grid of horizontal output tube V11 through the 2.2-megohm and 100-ohm resistors.

A positive voltage (+270V) goes to the HV adjust control through a 33K resistor divider. The control adjusts the amount of positive voltage applied to the VDR to counteract the negative voltage from capacitor C153 and provide a way to set the drive bias on the horizontal output tube to a specific value. increases or decreases due to a HV change, the driver bias to This sets up the HV regulation point. If the nominal 1500V pulse the horizontal output tube will increase or decrease, varying the tube gain of V11. Thus, the HV will stay constant at a point which set by the HV adjust control.

Voltage Dependent Resistors: High impedance and nonlinearity are the characteristics of a VDR. The kind of VDR used in this high-voltage regulator circuit is such that an increase in voltage causes a decrease in resistance, and a decrease in voltage causes an increase in resistance, which means that it works in an inversely proportional manner.

The lower waveform (A) in Fig. 7-9 depicts the 1500-volt pulse fed to capacitor C153. The upper waveform (A) represents the charging current of the capacitor. As the positive pulse from the sweep transformer nears point A, the charging current of capacitor C153 is approaching A′. During the negative protion of the applied pulse (point B, at lower left), the resistance of the VDR has increased and the very low amplitude of this negative portion affects the capacitor charging current (point B′, at right). This means that the *rectification* of the negative portion of this pulse has been accomplished, the same as with a diode or rectifier. Thus, as the capacitor is charging, it has a low resistance path through which to charge, but a higher resistance path through which to discharge. For no bias on control grid of sweep output tube V11, check for +140V on the screen grid (pins 3 and 11). Check the B+ supply for 270V. Check the sweep transformer windings for 270V B+ (damper tube V13 may be open or otherwise faulty). If the preceding checks are good, the loss of drive is probably due to some trouble in the horizontal oscillator circuit.

If there is insufficient bias on control grid of V11, again check the screen voltage for +140V. Check the action of the 6U10 horizontal oscillator and discharge stages for proper drive signal. The VDR (R110) may be defective or other

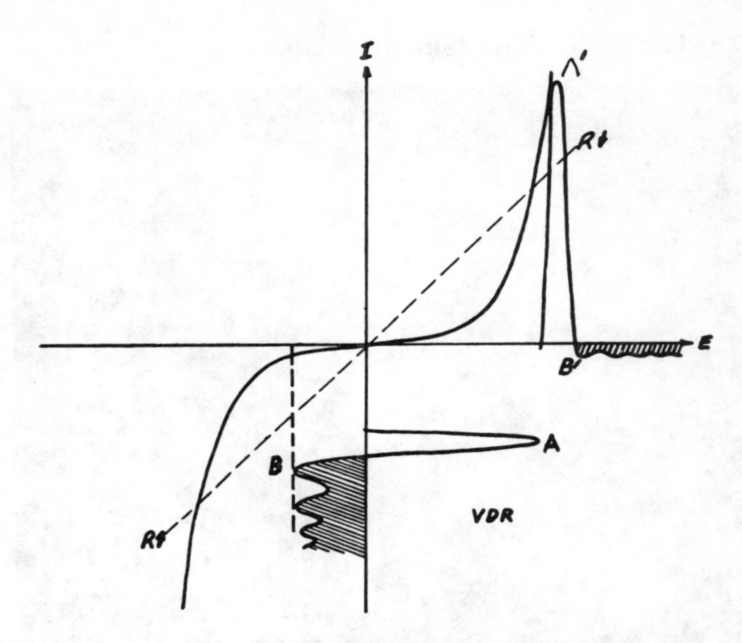

Fig. 7-9. VDR voltage and current chart.

components in HV adjust circuit may be faulty. The best check for a VDR is to replace it with a known good one. Measure the 2.2M resistor from the VDR to the control grid of V11 (pin 5), as it may have opened up or increased in value. Check the 120 pF capacitor (C153) for leakage or value change. And with your scope, check for the positive 1500V pulse at the orange lead of the sweep transformer.

For loss of high voltage check for +720V of the filtered boost. A correct boost voltage usually indicates that the sweep system is working correctly. Check the HV rectifier, and if it is OK, suspect the CRT. To make this check, remove the HV lead from anode of picture tube, turn on the set, and measure the lead with an HV probe. If you now have adequate HV, this would indicate a defective CRT or improper bias voltages on the picture tube control circuits.

In this set, loss of high voltage regulation was caused by an open 33K resistor (R111) that supplies +270 volts to HV adjust control R85. A faulty VDR could cause the same symptoms.

ZENITH 20Y1C50 CHASSIS

Picture Symptom: The set would snap, crackle, and pop when first turned on. The picture was very bright but a little blurred. The HV voltage measured at the CRT anode cap was over 30 kV. The picture would bloom as the brightness control was varied, which indicated the set had poor high-voltage regulation. Look for trouble in the high-voltage regulator circuit.

Circuit Operation: This chassis used the pulse-type HV regulator circuit shown in Fig. 7-10, with a 6HS5 HV regulator tube. The pulse regulator is a much more efficient means of keeping the HV on an even keel than a shunt-regulated HV system. The shunt system has continuous dissipation, while the pulse system has the loading taking place in the primary windings.

The pulse system is based on the constant loading and unloading of the sweep transformer primary winding, which reflects this change in the reduced load during low brightness. Of course, this all happens during the retrace time, and as a result the timing of the pulses are changed.

The pulse regulator tube is controlled by two actions. First, a variable DC voltage from the bootstrap voltage is used to control conduction of the regulator tube. Then, narrow timing pulses of constant width and amplitude from the horizontal oscillator let the regulator conduct only during retrace time. In the simplified schematic in Fig. 7-10, the 870V boost voltage is fed to HV adjustment control R76 via VDR R75. The voltage dependent resistor is a non-linear high impedance device and is used to keep the bias on the regulator tube V21 on an even keel.

The high voltage for the CRT and boost voltage varies in the same direction and are inversely dependent on the brightness variation of the picture. If brightness increases, the CRT anode and boost voltage goes down, and vice versa. This is why the boost voltage is used to control the bias of the regulator tube. The timing pulses ride on top of the DC voltage developed from the boost voltage, and the amount of regulator tube conduction is controlled by how far the timing pulses exceed the predetermined level of grid voltage. If the DC level set by HV adjustment R76 is low enough, tube V21 may completely cutoff. The variation of the boost voltage and timing pulses varies the current through the regulator tube, and as the current through the regulator tube is varied, the

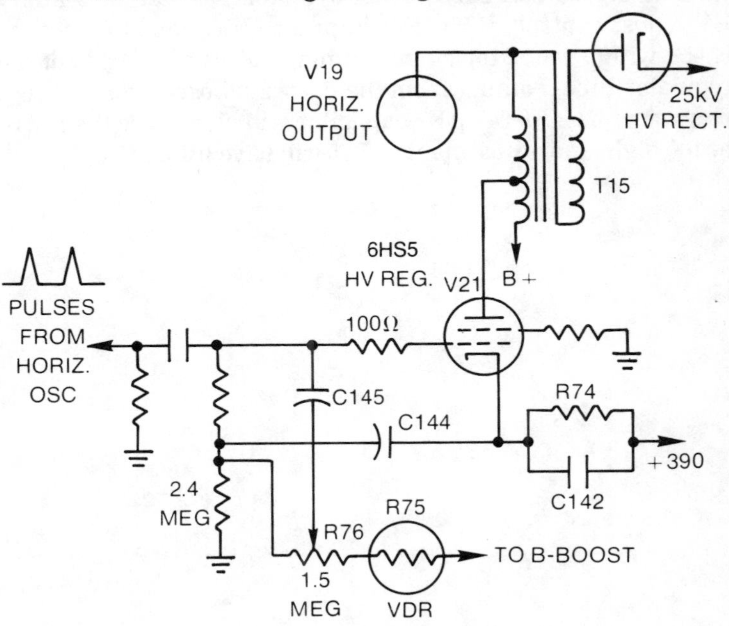

Fig. 7-10. Pulse HV regulator circuit.

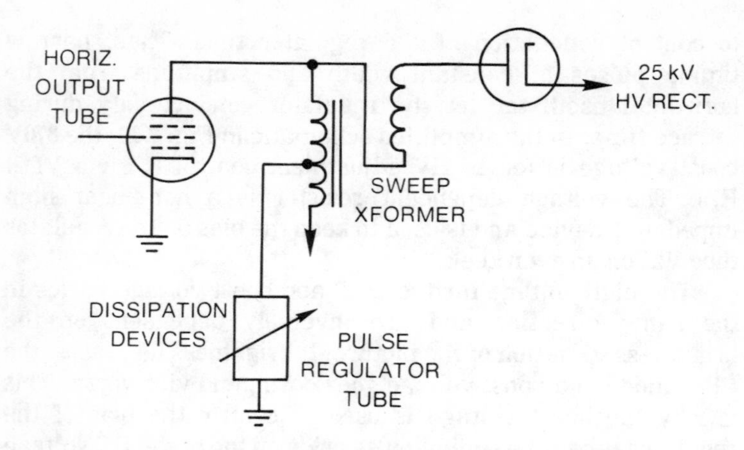

Fig. 7-11. Simplified pulse-regulator diagram.

regulator tube presents a varying load to the primary of the HV sweep transformer. Thus, the overall tuning of the transformer is changed and the HV is regulated.

Probable Cause: This very high voltage with no control was caused by a defective VDR. An open 220Ω R74 in the cathode of V21 will cause same symptom, as well as a faulty HV adjust control (R76). For improper HV regulation action, check with a scope for correct timing pulses at the grid of the regulator tube, coming from the horizontal oscillator. A leaky or shorted C144 (0.047 μF) capacitor will also cause the HV to be too high, and adjustment of R76 will have little effect.

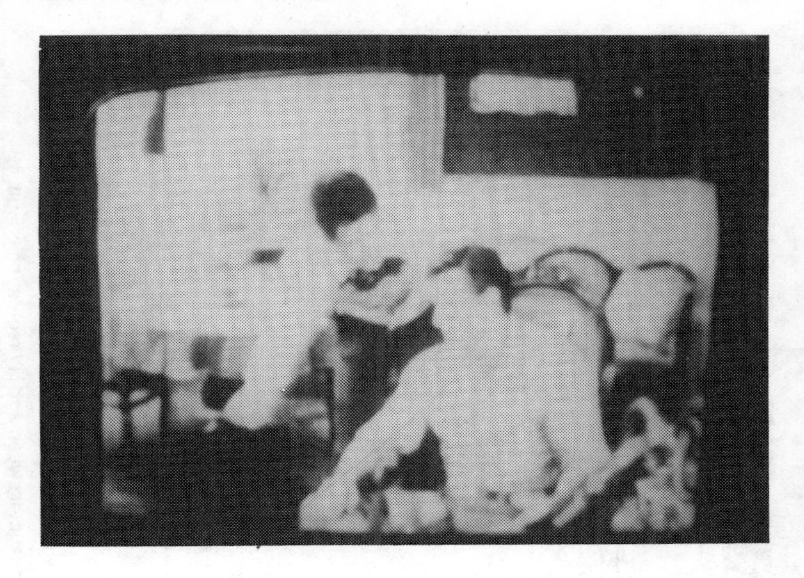

Picture Symptom: Picture pulled in from both sides. The high voltage at the CRT anode was low, and HV regulator adjustment R85 had no effect on picture or the HV. Look for trouble in the horizontal sweep output or feedback HV regulation system.

Circuit Operation: In this chassis, high voltage is regulated by automatically controlling the grid voltage of the horizontal output tube. This type system is much more efficient than a shunt-regulation system because all of the current generated is used by the picture tube.

The high voltage (21.5 kV) regulation is accomplished by a pulse-type feedback control circuit (Fig. 7-12) that automatically controls the bias (drive) voltage on the grid of the horizontal output tube. As an example, if the CRT beam current increases, causing an increase in the load on the sweep transformer, the high voltage will decrease, due to the decrease in the high-voltage pulse. The decrease in pulse amplitude will also cause a decrease in the 200V pulse coupled to regulator adjust control R85 through capacitor C153. Essentially, the capacitor charges through diode X17 during the pulse time (according to pulse amplitude) and discharges through the regulator adjust control. The amount of negative

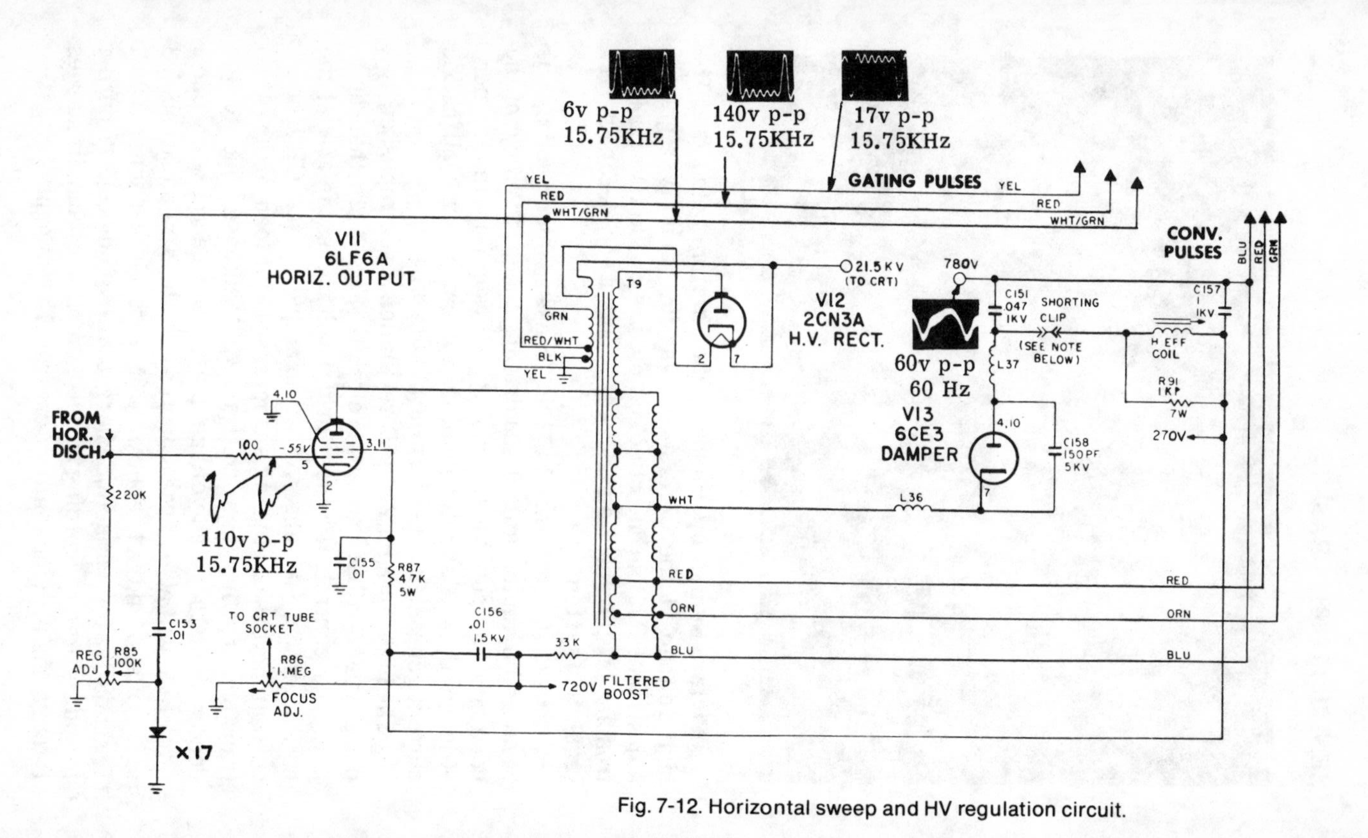

Fig. 7-12. Horizontal sweep and HV regulation circuit.

voltage across the regulator adjust control will be less with a decreased pulse and therefore lower the grid bias of horizontal output tube V11 (6LF6A). In turn, this will cause greater conduction of the output tube, which tends to increase the high voltage pulse, and thus the high voltage. In this chassis the regulator adjust control is set normally for 21.5 kV of high voltage, or just at the point of grid current flow in the horizontal output tube. The operating point for a negative grid-voltage regulation is adjusted by regulator adjust control R85.

Probable Cause: This pulled-in picture was caused by leakage in C153, the pulse feedback capacitor. Of course, this same picture symptom could be caused by an improper horizontal drive signal or a weak horizontal output tube (V11).

Each time the color chassis is serviced, the regulation system should be checked for proper operation. Measure the high voltage while varying the control setting. If the voltage can be set to the proper value, and varies with different high-voltage control settings, the system is operating properly.

Vertical Lines and Bars

In this section an attempt has been made to show how to link certain characteristic disorders to specific circuit faults. Although the listings are by no means all-inclusive, you will undoubtedly come to relate certain recurring symptoms with key trouble areas. For example, some narrow vertical lines can be caused by a high-voltage arc somewhere in the horizontal sweep and HV system. When an arc occurs across an insulator such as fiber or Bakelite, a carbon deposit forms. And the carbon deposit is conductive, allowing a small current to flow and encouraging further arcing in the same place.

In general, vertical bars can be traced to the horizontal circuits and the vertical circuits. Color bars can be traced to the 3.58 MHz oscillator stage. Some type of oscillation within the chroma circuits may produce narrow vertical color stripes. For vertical color bar troubles, refer to the chapter on color symptoms.

GENERAL ELECTRIC YA CHASSIS

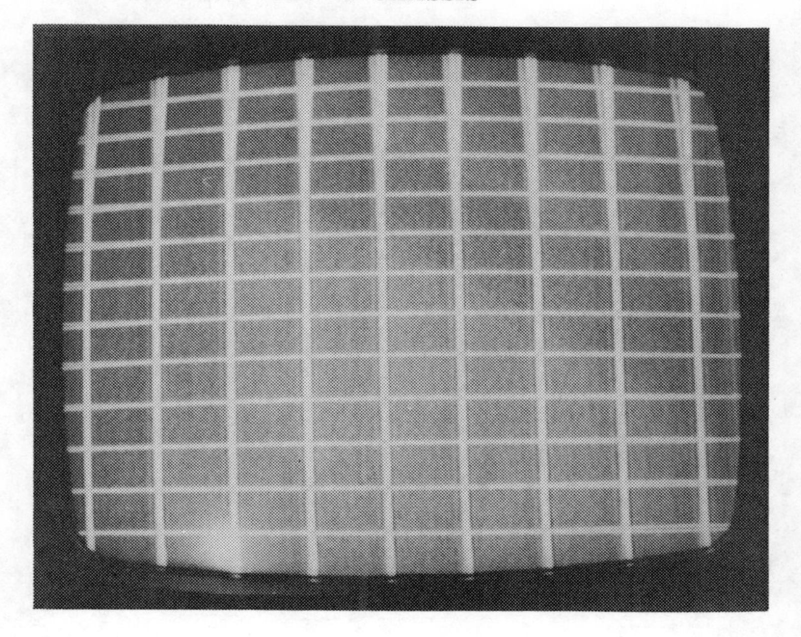

Picture Symptom: With a crosshatch generator connected to the receiver, the top vertical lines could not be converged properly. Look for trouble in the vertical dynamic convergence control circuits.

Circuit Operation: The vertical input terminals, pins 2 and 4 in Fig. 8-1, are in with the vertical deflection windings of the yoke. The vertical-rate convergence current is shaped mainly by the bridge consisting of diodes Y802 through Y806, and this current is added to the horizontal-rate convergence current from the control winding of saturable reactor T815. R808 is used to adjust the convergence near the top of the raster, when diodes Y802, Y803, and Y805 are conducting. R807 is used to adjust the convergence near the bottom of the raster, when diodes Y804 and Y806 are conducting. R810 reduces the power dissipated in R807 and R808, while at the same time limiting their range of control. Extra diode Y803 and resistor R805 are used to improve the convergence current waveshape near the top of the raster.

During vertical retrace time, due to the inductance of the vertical deflection winding, a negative pulse appears at pin 4.

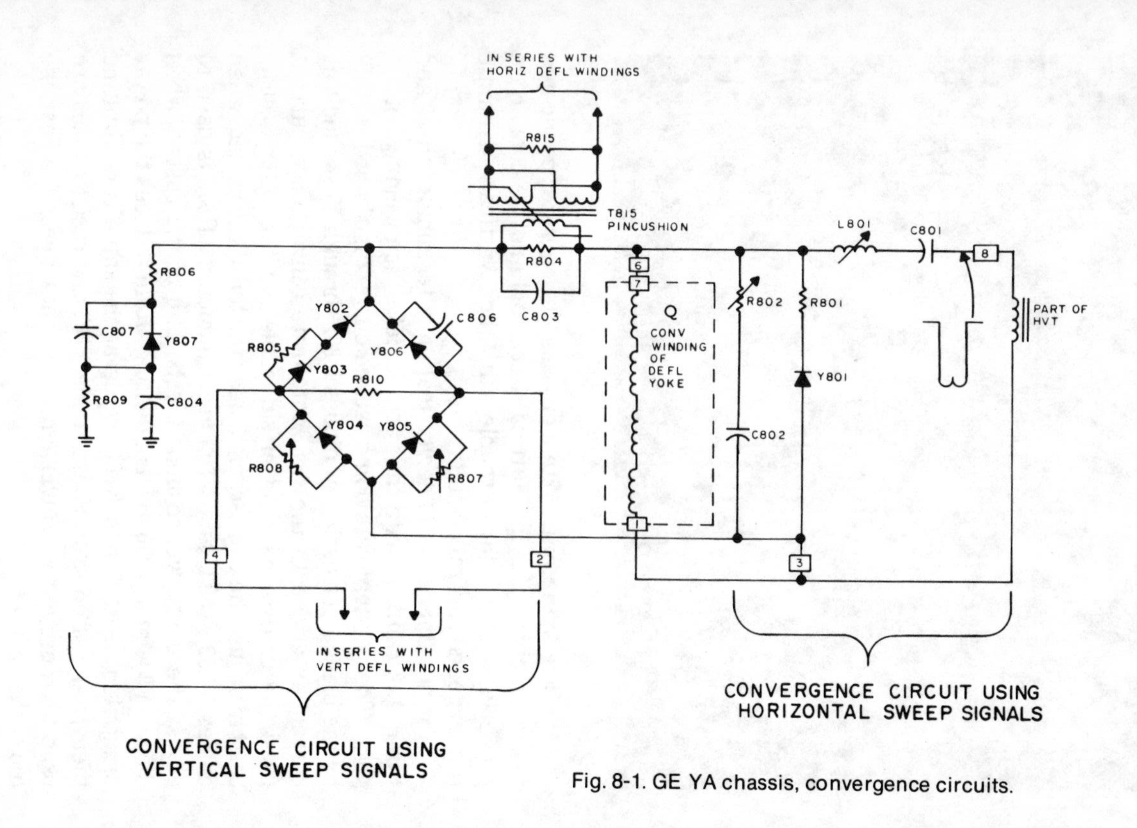

Fig. 8-1. GE YA chassis, convergence circuits.

Diode Y804 conducts through the convergence winding, the control winding of T815, R806, and Y807 during the vertical retrace interval, charging capacitor C804. Resistor R809 partially discharges C804 during the trace period. The current added by diode Y807 improves the convergence at the raster top.

Resistor R804 and capacitor C803 are connected across the control winding of T815 in order to suppress or damp any horizontal-rate pulses induced from current flowing in the controlled secondary winding.

Probable Cause: The vertical convergence problem here was caused by an open R805 control. Faulty diodes Y803 or Y802, or an open Y807 will cause the same problem.

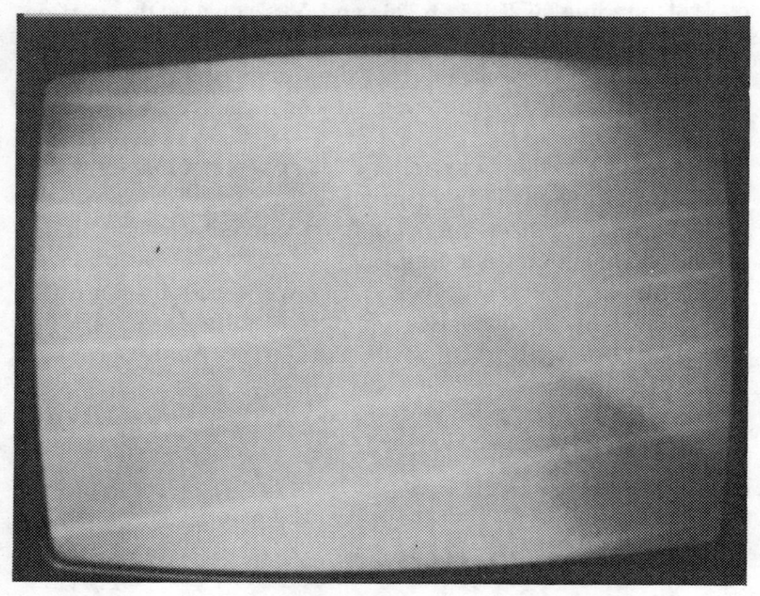

Picture Symptom: Picture very bright and blurred, with no video. There may be retrace lines. This picture problem could be caused by a fault in the video direct-coupled amplifiers or a faulty chroma demodulator (IC501); however, this particular fault was in the blanking and brightness-limiting stages located in the cathode circuits of the CRT.

Circuit Operation: Now for a quick rundown of how the blanking and brightness limiting circuits in Fig. 8-2 function. Negative-going horizontal and vertical retrace pulses are coupled to the base of Q111, a PNP transistor, which acts as a switch, being either full on or off. The collector signal is a series of horizontal and vertical blanking pulses, whose amplitude is 20V P-P. These pulses are coupled to the CRT cathodes through C166 and the CRT drive controls. Y107 protects Q111 from CRT arcs and sparks.

The ground return path for the CRT cathodes is through R197 and brightness-limiting transistor Q113. Base voltage for Q133 is set by R186 and R187, keeping the transistor saturated for normal operation. It will remain in saturation until the CRT beam current reaches its design limits. The beam current flows through R197, and as it increases, the emitter voltage of

Q113 becomes more positive and decreases the forward bias on the base–emitter junction. When this occurs, Q113 is not in saturation and its collector-emitter voltage increases with beam current. Q113 is then functioning as a variable resistor in series with the CRT cathodes, and consequently limits the beam current.

Probable Cause: This picture symptom was due to a loss of the +135V power supply at resistor R187. An open R408 (220Ω 2W) resistor in the power supply caused loss supply voltage at this point, and drastically upset the bias on the CRT. An open brightness-limiter transistor (Q113) would result in a dark screen. A defective vertical blanking transistor (Q112) will cause vertical retrace lines across the screen.

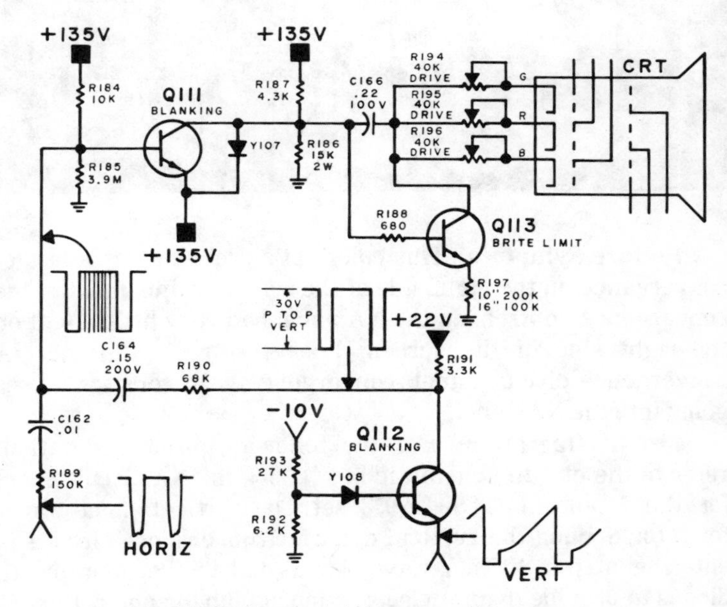

Fig. 8-2. GE JA chassis, brightness and blanking circuits.

MOTOROLA HTS-931/938 COLOR CHASSIS (QUASAR)

Picture Symptom: This color set's problem was a loss of convergence on the right side of the screen. Adjustment of the convergence controls on the NA panel had very little effect on the right side of the screen. Look for trouble in the NA convergence circuit panel, convergence yoke coils, or pulses going into the NA panel.

Circuit Operation: For the following circuit operation, refer to the circuit schematic in Fig. 8-3, the NA circuit panel for the Motorola HTS-931/938 set. The vertical input waveform for driving the vertical dynamic convergence circuits is entering at pin 6. This waveform is fed to the appropriate diodes to provide dynamic correction at both top and bottom of the screen for all three colors.

At the beginning of the vertical sweep (time T1 in Fig. 8-4A), the waveform is maximum in the positive direction. Diode D5 now conducts and this positive voltage appears across R2. This positive voltage, in turn, is fed through R4 to the red and green convergence coils (L702/A and L703/A). Diodes D6 and D7 are reverse biased with this positive voltage applied and look like open circuits. However, diodes D10 and D11 conduct to provide a return path to ground for that end of the convergence coils. The resulting current through the convergence coils is illustrated in Fig. 8-4C.

The adjustment control R2 increases or decreases the amplitude of the waveform applied to both coils. This results in separation or convergence (bending) of the top red/green vertical lines as viewed on the screen with a crosshatch pattern. An adjustment of R4 shifts the amount of current between the two coils and results in separation or convergence of the top horizontal lines. As the electron scan moves vertically toward screen center, less correction is required, so the vertical driving waveform reduces to zero when no correction is needed (time T2 in Fig. 8-4A).

As the scan moves toward the bottom of the screen, an increasing negative voltage is applied through D4 and then appears across R-7. This voltage is then fed through R3 to the red and green convergence coils. The other end of both coils is now returned to ground through D6 and D7. Notice that the waveform in A of Fig. 8-4 is now negative, but the polarity of

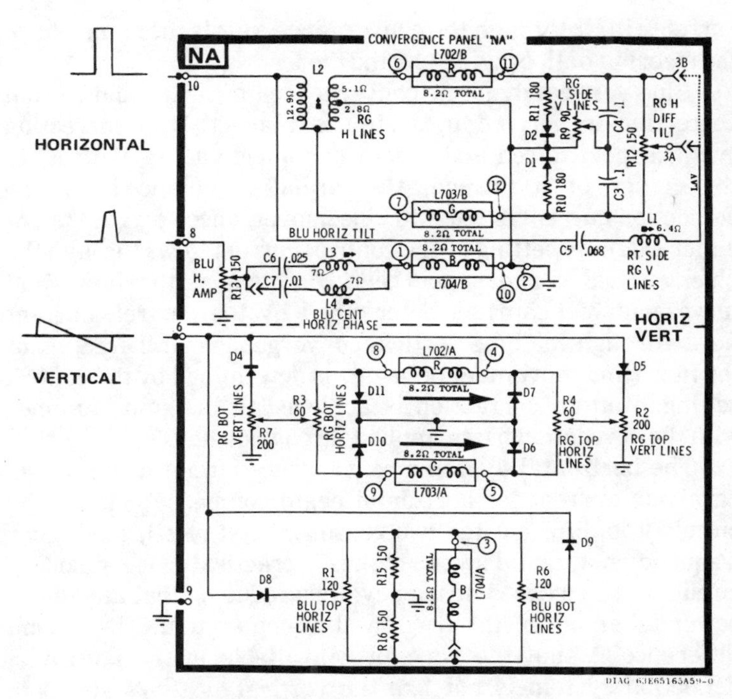

Fig. 8-3. Quasar HTS-931/938, NA circuit panel.

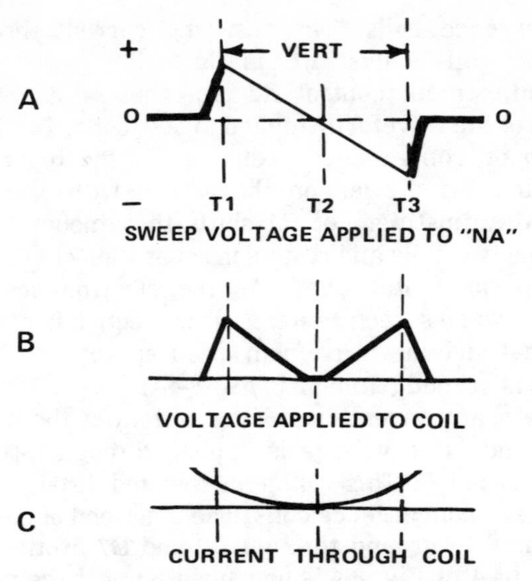

Fig. 8-4. Convergence current and voltage waveforms for vertical dynamic convergence.

current (in C) through the convergence coils is the same due to the steering path provided by the diodes.

Blue vertical dynamic convergence for the top and bottom correction is derived in a similar manner. Top correction (with respect to red and green) is applied via D8. With R1 in the center of its range, the circuit is balanced and no correction current is fed into blue convergence coil L704/A. At either extreme setting of the control, current flows through the blue vertical convergence coil. Direction of the flow (and movement of beam) is determined by the control, current flows through the blue vertical convergence coil. Direction of the flow (and movement of beam) is determined by the control setting. Bottom correction is achieved in the same manner with the correction being applied through diode D9.

The horizontal dynamic convergence circuit also needs a parabolic current for horizontal beam compensation, but the circuitry is different for two reasons. First of all, the higher frequency horizontal sweep makes it practical to use resonant circuits for increased efficiency, thus reducing that amount of power taken out of the horizontal sweep system. The second difference is that the waveform available in the horizontal deflection system is not like the vertical system's sawtooth voltage, but is instead a short pulse. Circuitry must then be used to shape the pulse to the proper waveform to cause a parabolic current to flow in the convergence coils.

Variable inductors plus the impedance of the convergence coils themselves change the pulse from the horizontal sweep output transformer into a sawtooth voltage, which then appears across the convergence coil. The variable inductance makes it possible to vary the voltage across the convergence coil, and current through the coil then varies with the voltage to provide current control. These voltages, currents, and pulses are illustrated in Fig. 8-5.

Actually, the horizontal coil must resonate at a frequency somewhat lower than the scan frequency. It is resonated so that it completes one-half cycle in the time it takes the beam to scan from the left to the right side of the CRT.

Probable Cause: Most convergence circuits do not give too much trouble, but when they do, the fault may be hard to find. A good way to start tracking down trouble is to attempt to make all of the convergence adjustments. As you turn each control, see if it will produce the proper action. If you find a control that has little or no effect, then the prime suspect would be the circuit in which this circuit control operates.

The loss of picture convergence on the right side was caused by an open L1 coil, the right side red and green vertical line adjustment coil.

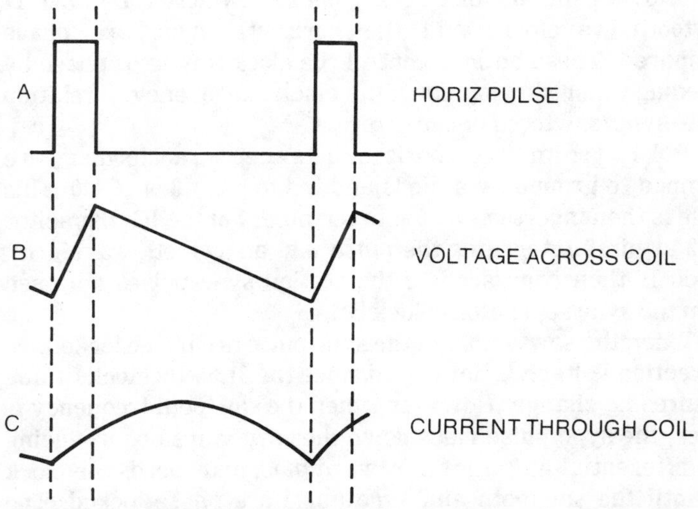

Fig. 8-5. Comparison waveforms of horizontal dynamic convergence.

Picture Symptom: Vertical roll and/or picture jitter. Look for trouble in the sync processor circuits (IC400) or the vertical /horizontal countdown system (IC300).

Circuit Operation: The IC400 chip (Fig. 8-6) contains a 31.5 kHz clock (block E) that is phase locked to the horizontal sync signal by a comparator circuit (block D), where the 15,750 Hz sawtooth waveform and the horizontal sync are phase compared. These actions control the clock timing or phase by speeding up or slowing down the clock's frequency in relation to the sync/sawtooth phase error.

Pulses from the horizontal sweep transformer are clamped to ground by a diode and fed to point 8 of IC400. This pulse is then inverted and fed to terminal 7 of the IC. Capacitor C422 at pin 7 integrates the pulse into a sawtooth waveform, which is then compared to the station sync pulses obtained from the sync separator (block C).

When the sawtooth and the sync pulse are phase locked, no correction voltage is developed since the 31.5 kHz clock timing requires no change. However, when the sawtooth frequency is lower, the sync pulse rides down the sawtooth slope, reducing the differential amplifier's forward bias, and speeds the clock up until the sawtooth and sync pulse are phase locked. The clock pulse shown in Fig. 8-7 was taken at pin 15 of IC400.

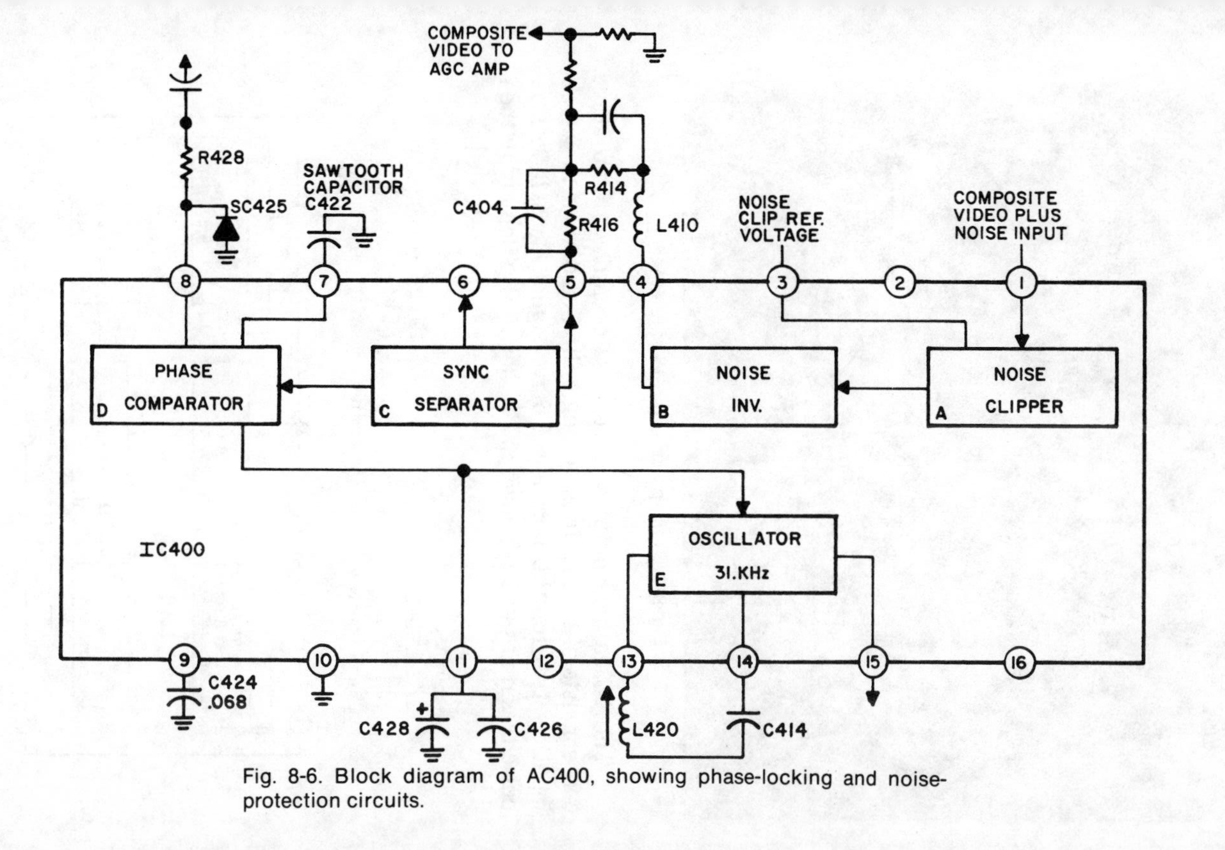

Fig. 8-6. Block diagram of AC400, showing phase-locking and noise-protection circuits.

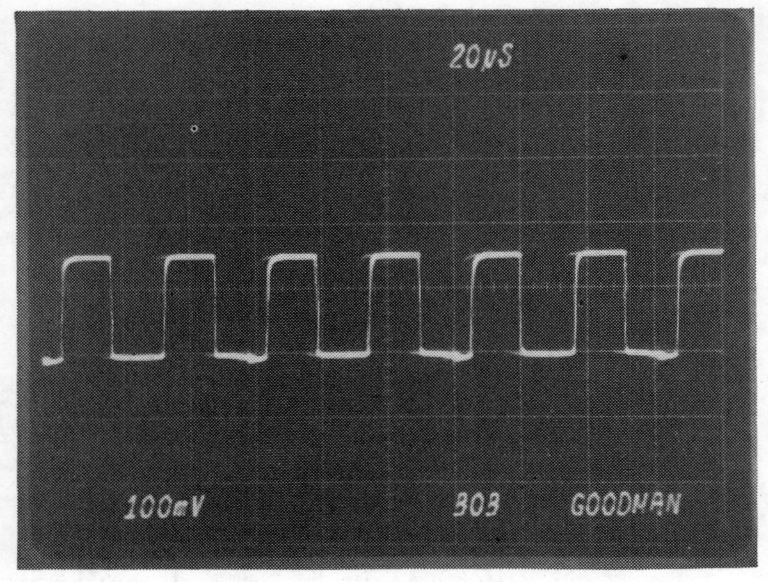

Fig. 8-7. Clock pulse at pin 15 of IC 400.

The IC300 horizontal and vertical countdown chip (Fig. 8-8) consists of six sections. A single flip-flop divides the clock input by two for the horizontal drive signal. This 15.75 kHz pulse taken at pin 2 is shown in Fig. 8-9. A flip-flop array then

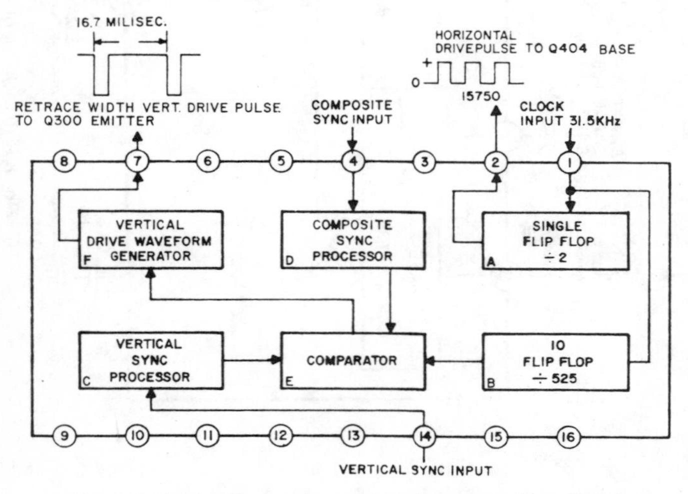

Fig. 8-8. Vertical countdown signal from pin 15 of IC300.

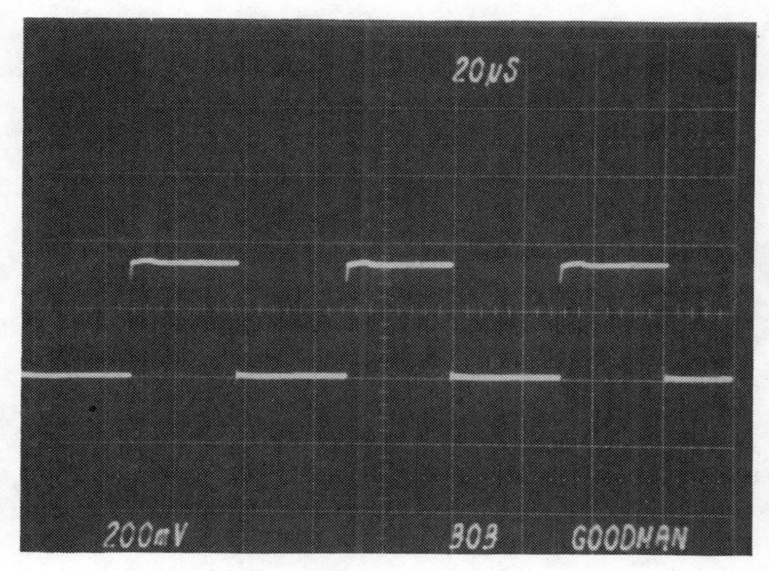

Fig. 8-9. The 15.75 kHz pulse at pin 2 of IC300.

divides the clock input by 525 to create the synchronizing signal shown in Fig. 8-10.

A composite sync processor checks the vertical sync pulse for the presence of equalizing pulses. A vertical sync processor

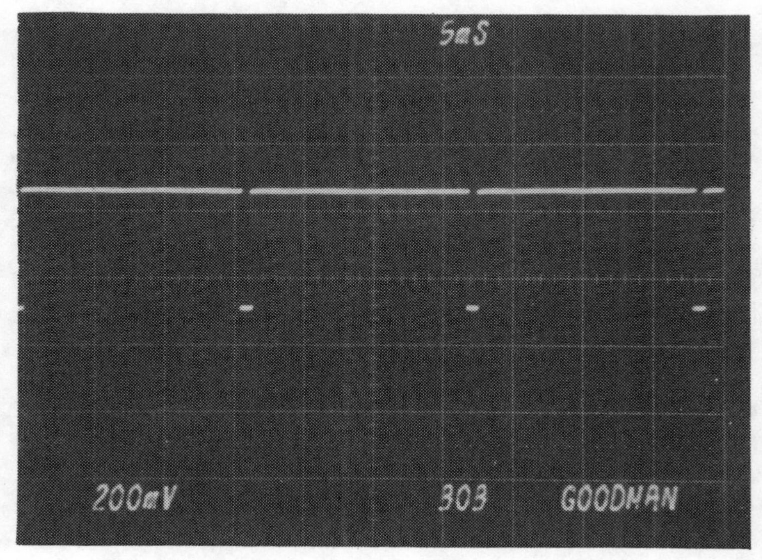

Fig. 8-10. Vertical sync input pulse at pin 14 of IC300.

clocks the vertical drive pulse for non-interlaced signals. And a comparator circuit produces mode-switching logic to control the vertical drive waveform generator that puts out a retrace-width vertical drive pulse.

In operation, the composite sync signal is sampled by the composite sync processor (block D) for the presence of equalizing pulses to determine if the signal is or is not interlaced. Block D and block E are the logic circuits that determine the appropriate vertical sync operating mode. While block A is a single flip-flop producing a clocked 15.75 MHz output for horizontal drive to make an accurately synchronized scan system.

Probable Cause: This vertical roll trouble was caused by a faulty IC300 chip, which may also cause a picture jitter symptom.

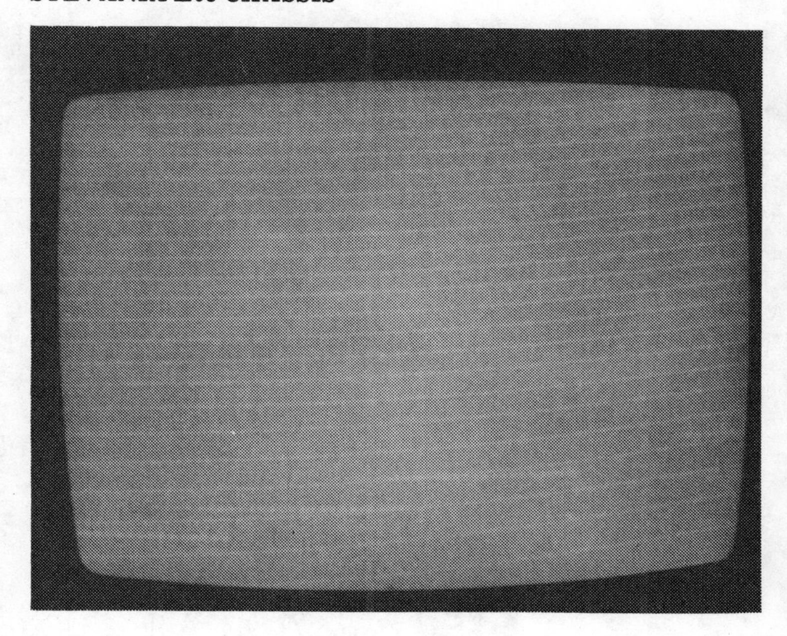

Picture Symptom: Thin white lines across the screen, which indicate vertical-retrace blanking problems. Look for trouble in the vertical blanking circuits.

Circuit Operation: Referring to Fig. 8-11, note that the vertical sweep pulse is polarized with the retrace spike negative-going. This pulse is coupled through C912 and SC910 from vertical output stage resistors R364 and R366 to emitter resistors R915 and R916 of Q904. Capacitor C912 places the AC waveform on the DC voltage found at the junction of C912 and SC910. Resistor R914, connected to the +23V supply, reverse-biases diode SC910, permitting the negative tips of the vertical sweep pulses to drive the emitter voltage of Q904 down from B+. The PNP transistor is turned off as its emitter voltage approaches its base voltage, producing a negative-going pulse across collector resistor R918.

Transistor Q908 is forward biased by the positive DC voltage across R918. So the negative-going spike reverse-biases the transistor and produces a positive-going pulse on its collector. Q908 is followed by Q910, a PNP transistor with its emitter connected through service switch

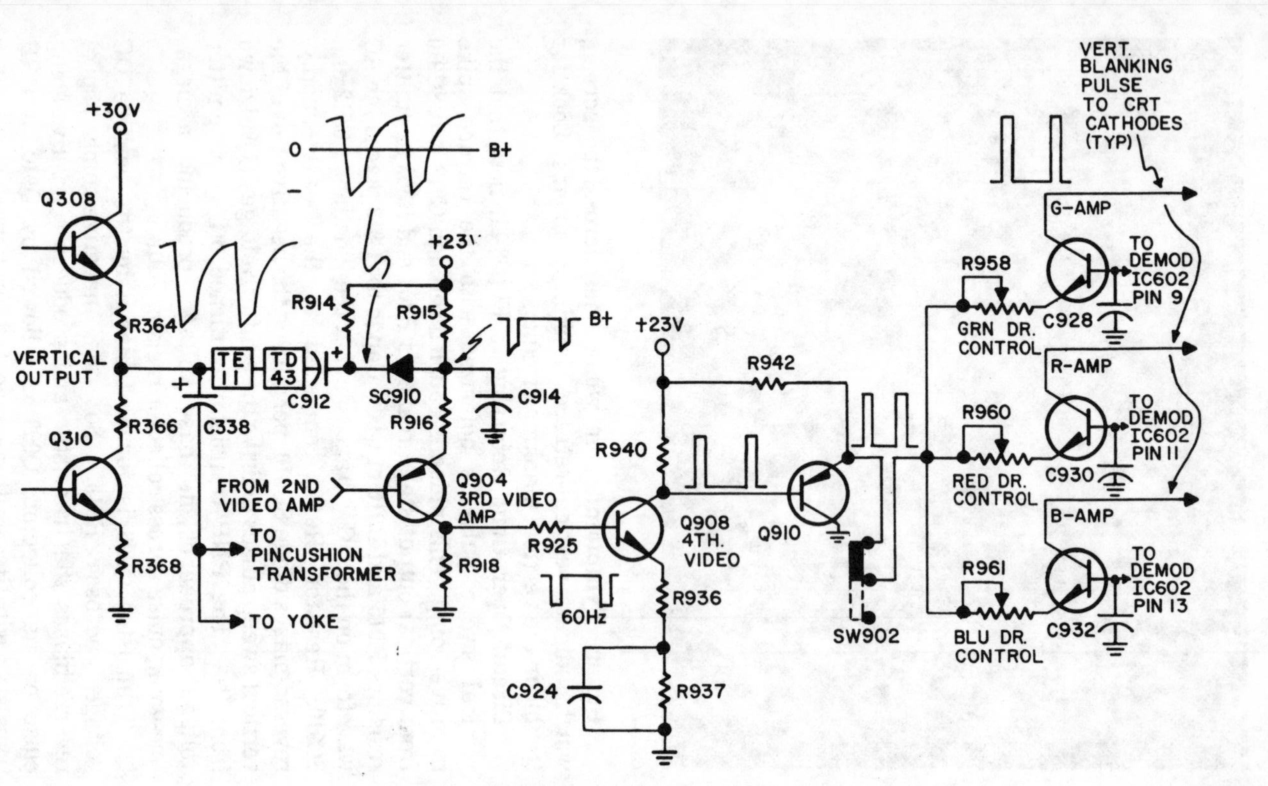

Fig. 8-11. Sylvania E05 chassis, vertical blanking circuit.

SW902 to the RGB drive controls. Video driver Q910 is an emitter follower, producing a noninverted positive pulse on its emitter. This pulse also travels through the emitter-follower RGB amplifiers and drives the CRT into blanking during vertical retrace time. This process eliminates vertical retrace lines from the raster.

Probable Cause: Using a scope to trace the vertical blanking pulse through this circuit, the loss of vertical blanking was found to be caused by an open C912 capacitor. A faulty SC910 diode could cause the same type picture symptoms.

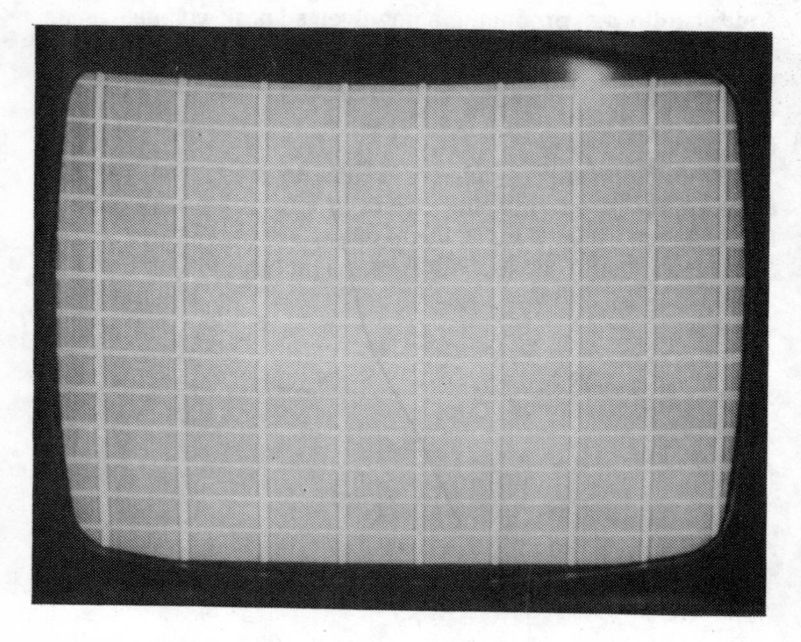

Picture Symptom: Top line of crosshatch pattern is bowed inward, and picture may be pulled in from top and bottom. Look for trouble in the pincushion correction circuitry.

Circuit Operation: Because the pincushion circuit has relatively few components, it does not develop too many problems. For quick diagnosis, the pincushion circuitry can be divided into two separate sections. One section contains the top and bottom pincushion correction while the second contains the side pincushion correction.

The pincushion circuit board in the Zenith 19EC45 chassis is mounted on top of the deflection yoke cradle and snaps on and off for servicing. The saturable-core reactors are mounted so that their operation is not influenced by any stray yoke field that might upset their operation and balance. The phase coil (Fig. 8-12) has an external ferrite sleeve that serves as a magnetic field to minimize interaction with other local fields.

The bottom and top pincushion circuitry consists primarily of tranformer T1301, Coil L1301, and capacitors C1303 and C1304. To determine if this circuit is functioning properly, perform the following procedures and checks. First,

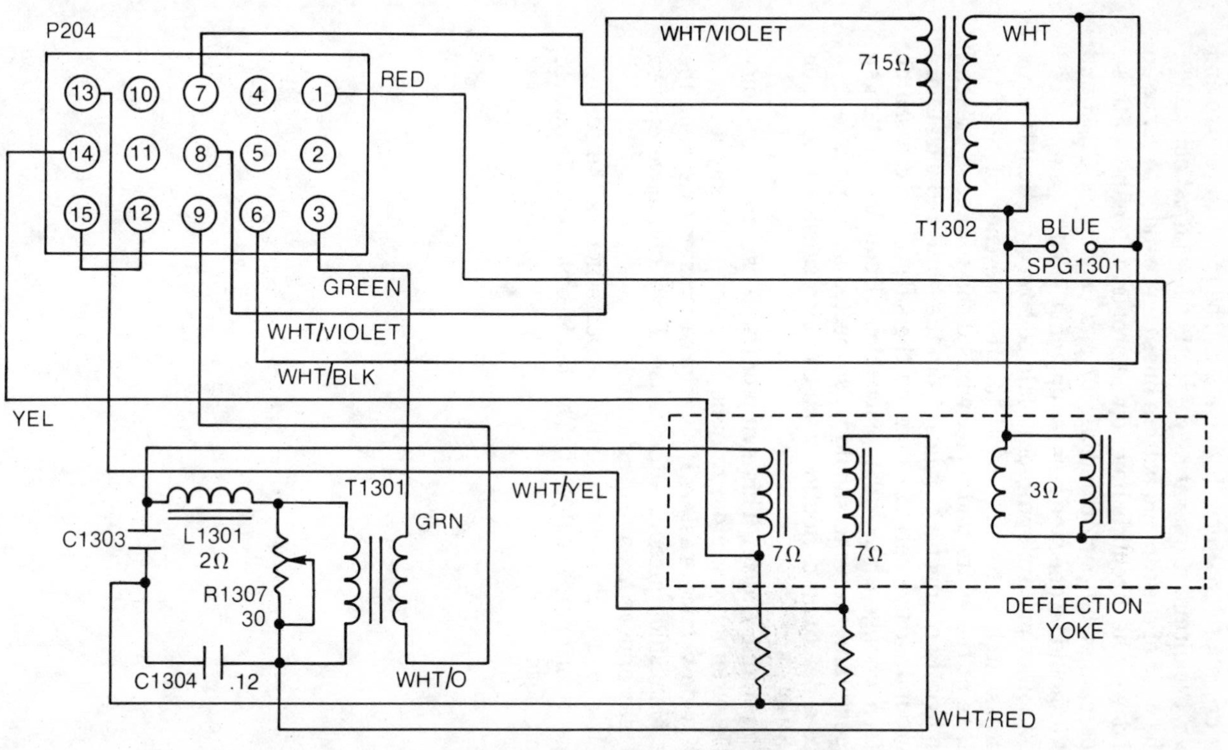

Fig. 8-12. Zenith 19EC45, pincushion circuit.

adjust R1307 and observe the center of the top and bottom lines on a crosshatch pattern. If the line can be made to bow up and down, the circuit is operating correctly. Now adjust coil L1301. If the ends of the top and bottom lines can be made to curve up and down, the circuit is functioning properly. If either of these steps cannot be performed, R1307 and capacitors C1303 and C1304 should be checked before replacing T1301.

Another picture symptom could be bowed vertical lines on a crosshatch pattern. Check for the presence of negative voltage at the white and violet wire. If the voltage is present, then briefly connect a clip lead from the white and violet wire to ground. If the vertical lines on the side of the crosshatch pattern change shape, the pincushion circuit is functioning properly and the yoke should be suspected as being defective since a defective deflection yoke can also cause a pincushion problem.

Probable Cause: This symptom was caused by a faulty C1304. Another symptom may be a picture with only a 1 or 2 inch wide vertical raster. If this narrow raster exists and the side pincushion circuit is the cause, T1302 may have opened and may appear to have overheated.

If a one inch horizontal line appears across the screen, suspect an open L1301 coil. The leads have been known to break loose.

ZENITH 19FC45 CHASSIS

Picture Symptom: Two bright, thin horizontal lines that move through the picture. This indicates a 120 Hz ripple on the B+ supply lines. Look for trouble in the power supply, such as an open filter capacitor.

Circuit Operation: Voltage regulating transformer T201 (Fig. 8-13) has a loosely coupled primary and secondary winding. The secondary winding of this VRT is tuned to 60 Hz with a 3.5 μF oil-filled capacitor (C2313). The secondary winding is the regulating element, for with the secondary resonating, the voltage may increase to a value where the core material will saturate, but any further increase in voltage is not possible. This regulating action might be compared with a double-sided zener diode arrangement.

The low-voltage power supply has a 177V AC input coupled to the primary of the VRT. The VRT inductively couples an AC voltage to its secondary where it is rectified by eight diodes and filtered to provide six DC supply voltages.

Probable Cause: These bright lines and distorted picture were caused by an open C214D, an 80 μF filter capacitor. On some sets this filter is a single unit mounted on the back side of the main chassis, while on other sets a four-section filter is mounted on the power supply subchassis. C214A and C214B are also prime suspects for similar picture symptom.

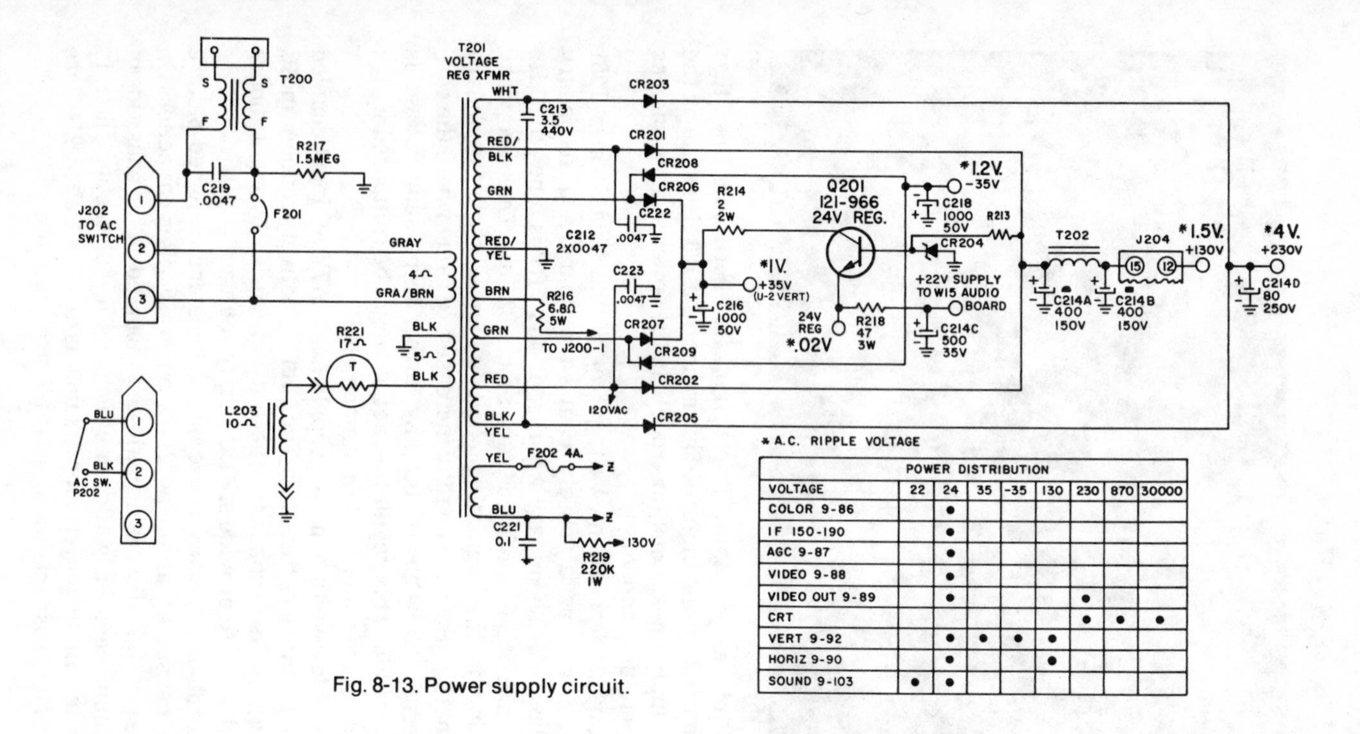

POWER DISTRIBUTION								
VOLTAGE	22	24	35	-35	130	230	870	30000
COLOR 9-86		●						
IF 150-190		●						
AGC 9-87		●						
VIDEO 9-88		●						
VIDEO OUT 9-89		●				●		
CRT						●	●	●
VERT 9-92		●	●	●	●			
HORIZ 9-90		●			●			
SOUND 9-103	●	●						

Fig. 8-13. Power supply circuit.

ZENITH 20X1C38 CHASSIS

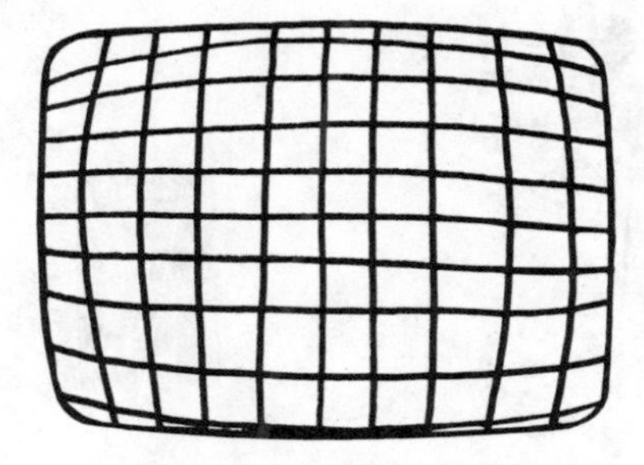

Picture Symptom: The crosshatch signal pattern had a barrel-shaped appearance, indicating too much pincushion correction. Look for trouble in the pincushion correction circuit.

Circuit Operation: On all color receivers that use the newer, rectangular, flat-faced picture tubes, you will always find that some type of pincushion correction circuit is used. The reason for this is that a normal deflection yoke develops a uniform magnetic field within its aperture when a linear current is flowing through the horizontal and vertical windings. On a picture tube with a spherical face plate, a rectangular raster will be produced on the screen. But a picture tube with a flat face will produce a raster with pincushion distortion (the opposite of barrel distortion). A failure of the correction circuitry can thus produce either pincushion or barrel distortion.

Refer to the pincushion correction circuit in Fig. 8-14. The 780V unfiltered boost (a 15 kHz parabolic waveform) is coupled to the pincushion transformer T5 through capacitor C85 and also to the control grid (pin 2) of correction tube V2B. A 60 Hz sawtooth voltage is also coupled to the control grid, which varies the signal amplitude or gain of the tube at the 60 Hz sawtooth rate, adding the two parabolic signals at the plate of the tube to form the required pincushon correction signal. The correction signal at the plate (pin 3) has the appearance of a bow tie, as shown in Fig. 8-15. The resulting parabolic plate signals are coupled through transformer T5 to the deflection yoke for top and bottom pincushion correction. Transformer T5 is adjustable for optimum correction.

The 60 Hz vertical sweep waveform is coupled to the cathode (pin 4) of the correcton tube to help achieve the

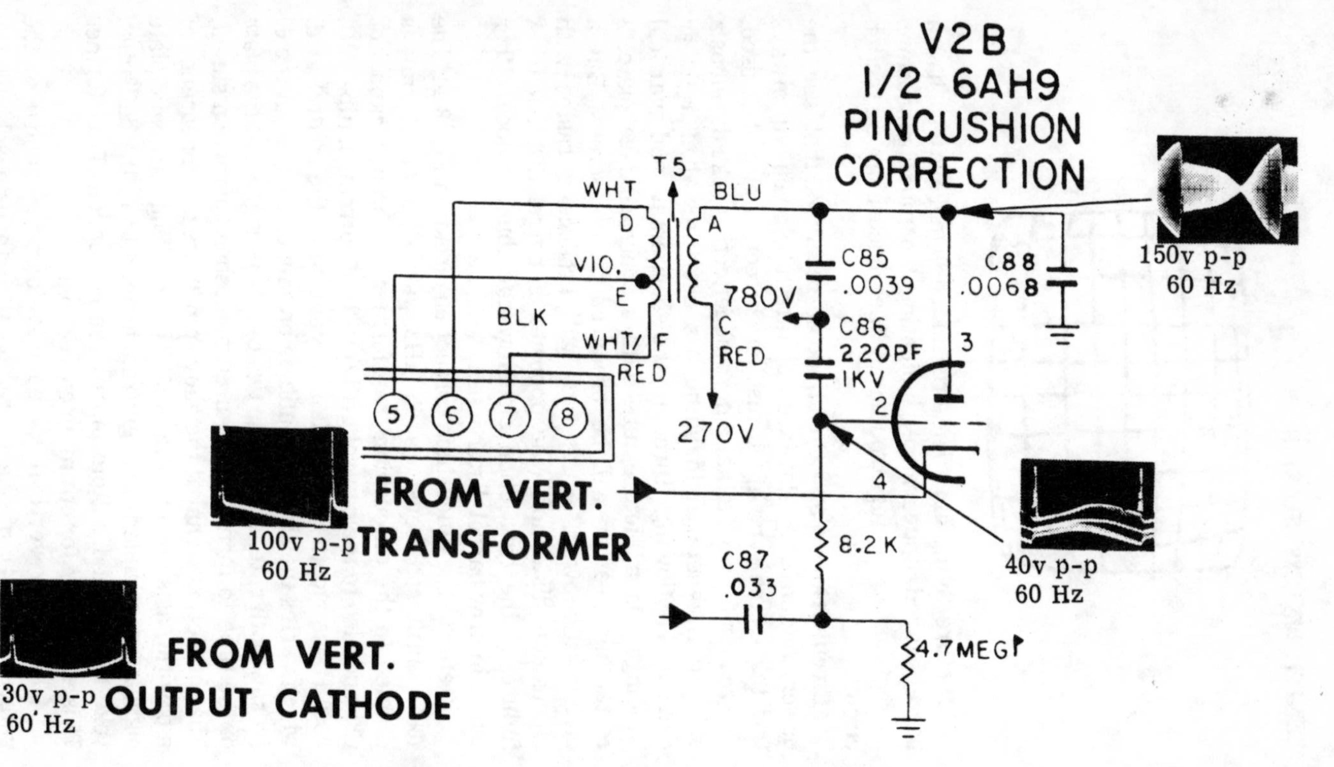

Fig. 8-14. Pincushion correction circuit, Zenith 20X1C38.

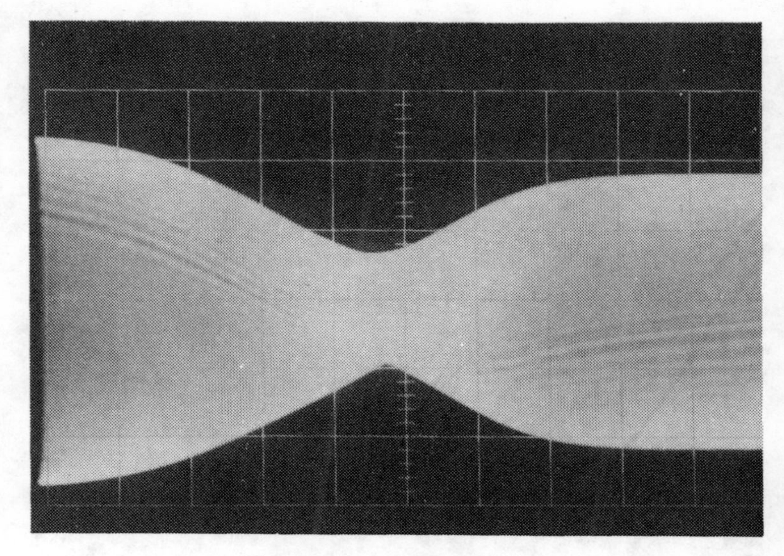

Fig. 8-15. Bow-tie scope pattern.

required shape of the output waveform throughout the vertical sweep interval.

Probable Cause: This picture symptom was caused by an open coupling capacitor, C85. To isolate a fault in this circuit, use a scope to trace out the vertical and horizontal signal correction waveforms.

ZENITH 20Y1C50 CHASSIS

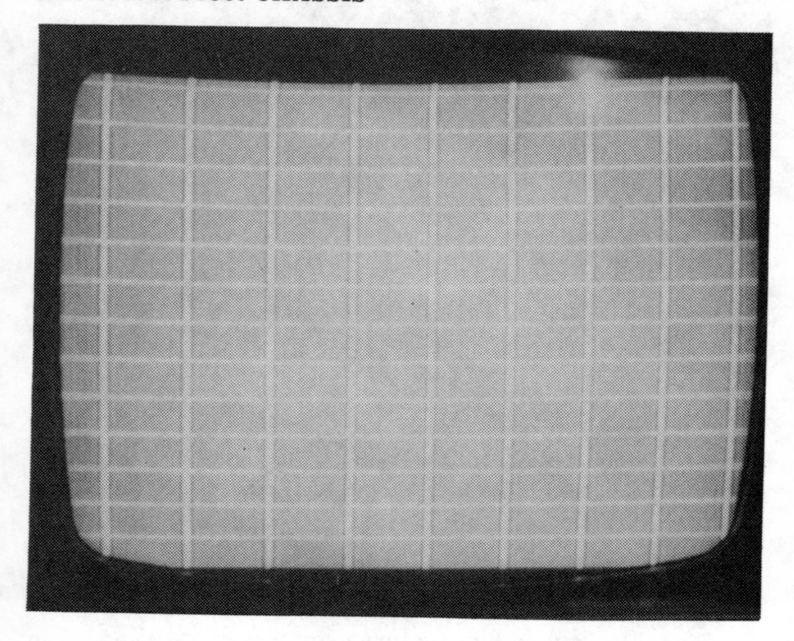

Picture Symptom: A pincushion effect at top of picture could not be corrected by adjusting the pincushion. Look for trouble in pincushion correction circuit.

Circuit Operation: Refer to the tube type pincushion correction circuit stage in Fig. 8-16. Some unfiltered boost voltage (horizontal parabola) is coupled through a 0.0012 μF capacitor to the pincushion transformer and on to the deflection yoke. Without the tube in the circuit, this voltage would produce maximum pincushion correction at the top of the raster, but would increase distortion at the bottom of the raster. If the boost voltage is coupled only through the 220 pF capacitor and the tube, the effect would be on 180-degree phase reversal of the signal, producing maximum pincushion correction at the bottom of the raster and increased distortion at the top. Thus, a tube is used in conjunction with the 60 Hz sawtooth voltage for complete control.

At the start of the 60 Hz sweep, the tube is at or near cutoff due to the cathode being positive due to the amplitude of the sawtooth voltage; thus, maximum correction is accomplished at the top of the raster. At the center of the sweep, the cathode is less positive, and the tube is conducting at a level that cancels the horizontal voltage; thus, minimum or zero pincushion correction occurs at the midsweep point. As the tube continues conduction, which becomes greater toward the bottom of the 60 Hz sweep, the horizontal voltage through the tube becomes greater and corrects the lower portion of the raster for pincushion.

The bead (L) on the grid-leak lead wire helps prevent oscillation, which can cause blobs on the screen. The voltage appearing at the transformer more closely resembles a sine wave due to the ringing action of the transformer when a pulse type voltage (horizontal parabola) is fed to it.

Probable Cause: The improper pincushion correction depicted was due to an open 0.0012 μF capacitor used to couple in the boost voltage. An open 1000Ω resistor in the cathode circuit would result in no correction, as would a bad tube. Do not overlook a defective pincushion correction transformer.

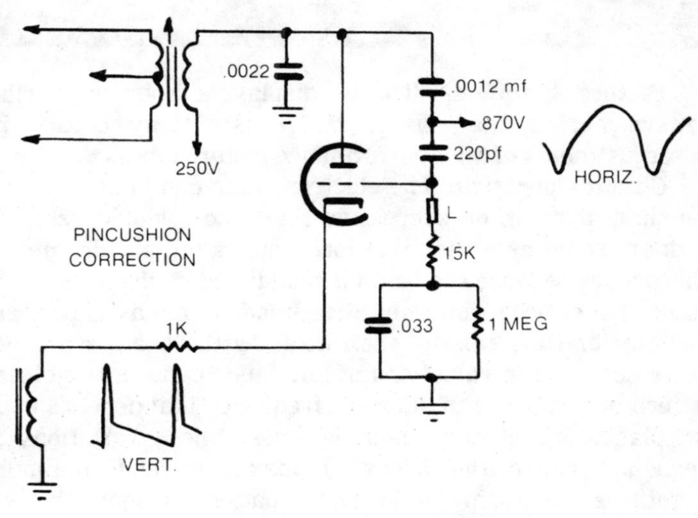

Fig. 8-16. Pincushion correction circuit.

Picture Symptoms: The set displays a moire or swirling pattern across the screen, which is usually caused by misadjustments of high-performance picture tubes.

Circuit Operation: The electron beam can miss a hole on the shadow mask on successive scan lines, which results in dark areas on the screen that form into patterns called *moire*. This occurs because the beam is modulated by the holes in the mask, generating an amplitude modulation as the beam partially or fully misses each hole. As the beam scans the entire screen, the variations in amplitude cause a brightness pattern describing a modulation frequency that depends upon the placement of the scanning lines. These variations in frequency cause the areas of maximum and minimum brightness to form a swirling pattern rather than a straight-line pattern.

Moire is usually more noticeable without video information present, but it should be considered a problem only if it is objectionable on programs at a normal viewing distance.

Probable Cause: The following list describes the adjustments and corrections that can be made to minimize the moire pattern.

- Adjust the anode voltage to the correct value. Too high a voltage narrows the beam and reduces spot size.
- Adjust the yoke by throwing the service switch and looking at the setup line. It should be exactly horizontal for minimum moire.
- Check the vertical centering and setup line. It should be exactly centered vertically.
- The height and linearity must be correct. Poor vertical linearity or improperly adjusted height will cause or exaggerate moire.
- Check the pincushion adjustment with a crosshatch pattern. Adjust the pincushion control through its full range, then stop at the minimum moire point.
- The screen should show good interlace. Adjust the vertical hold control while viewing the scanning lines.
- Also check and reset the CRT bias if it is needed.
- The final step is to focus for best highlight performance.

Moire should now be eliminated or at least reduced to an acceptable level. It is important to remember that any effort to reduce moire by increasing spot size only degrades the focus. Thus, care should be taken to insure that the spot size is not increased excessively.

Color Problems

Color problems usually appear as missing colors, wrong colors, rainbow patterns, or no color at all. Since the color section of the TV receiver is pretty much independent of the rest of the receiver (except for the picture tube), color problems can normally be isolated to the 3.58 MHz oscillator, burst amplifier, chroma demodulators and bandpass circuitry, convergence panels, automatic chroma control (ACC) systems, and the CRT control voltages and biases.

With the chroma or color control turned all the way down, the screen should produce a good black-and-white picture with no color blotches or color fringes around the objects in the picture. Color blotches indicate the need for degaussing and purity adjustments, while color fringes indicate poor convergence. Often the simple process of setting up the color TV for a good monochrome picture will pinpoint other color problems by the lack of response from various color-adjustment controls, thereby isolating the fault to the control's associated circuitry.

Complete absence of color (monochrome picture only) suggests that the 3.58 MHz oscillator, burst amplifier, color IF amplifiers, ACC, and chroma demodulator may be at fault. Wrong colors, on the other hand, may result from a phase (tint) error in the 3.58 MHz oscillator and the automatic control systems. The absence of one or more colors generally results from a fault in one of the chroma demodulators or color output amplifiers, or even a bad CRT.

An oscilloscope is a must for troubleshooting color circuitry, and a color vector scope is especially good at quickly pinpointing missing-color and tint problems when used in conjunction with a color-bar generator. A combination color-bar/crosshatch generator should be a part of every serviceman's outfit.

EXCESSIVE GREEN

Picture Symptom: Picture is all green, blurred, and has retrace lines. Most likely this fault will be found in the RGB video output stages. The color picture tube and bias circuits are also suspects.

Circuit Operation: In Fig. 9-1, transistors Q704, Q705, and Q706 constitute the RGB (red, green, blue) video output stages. Their DC bias is controlled by transistors Q707, Q708, and Q709, which function as series resistors from the emitter of each output stage to ground. By adjusting the bias control, the emitter voltage of each output stage can be independently controlled. This permits color balancing and compensation for CRT gun efficiency during black-and-white setup.

The video signal (luminance or "Y") is applied to the emitter of each output stage from video driver Q703. The video drive from each output transistor can be adjusted independently by its own drive control.

The color-difference signals (three signals: B − Y, R − Y, G − Y) from the demodulator section of IC400 are

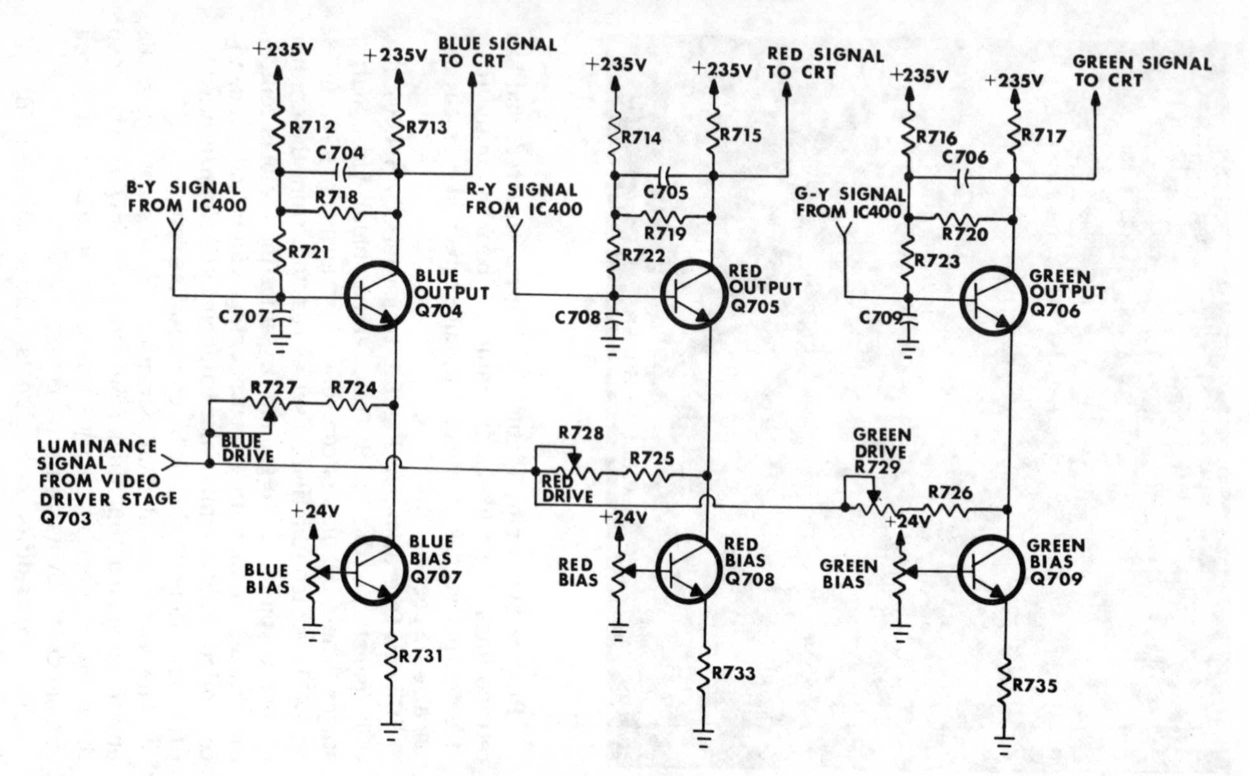

Fig. 9-1. Admiral M10 chassis, RGB output stages.

directly coupled to the bases of their respective output transistors.

With the demodulated color-difference signal applied to the base and the luminance signal applied to the emitter of each stage, matrixing of the two signals occurs within each transistor. The resulting blue, red, and green drive signals are coupled from the collectors of the output transistors to the respective CRT cathodes.

Capacitors C707, C708, and C709 are 3.58 MHz bypass units used to remove unwanted interference at the base elements. C704, C705, and C706 provide a degenerative feedback signal to each base element for stabilization. Resistors R718, R721, R719, R722, R720, and R723 provide DC feedback to minimize tracking changes during component aging.

Although this circuit description specifically covers the M10 chassis, the same principal of matrixing the color difference and luminance signals at the video output transistors is used in many other Admiral color chassis.

Probable Cause: The green screen was caused by an open R735 emitter resistor located in the green bias-transistor stage. An open Q706 or Q709 transistor could cause the same picture symptoms, as could a defective color picture tube.

Picture Symptom: This set had a good black-and-white picture, but had an intermittent loss of color. At times a color barber-pole effect could be seen, which indicated a color sync problem. Look for trouble in the color burst amplifier circuit.

Circuit Operation: Refer to color-burst amplifier circuit in Fig. 9-2. Burst amplifier Q703 is a PNP transistor operated in the common-emitter configuration. This transistor is reverse biased and is pulsed into conduction only during the horizontal retrace time.

The base of the burst amplifier receives two signals: the chroma information with burst present, and the horizontal control pulse. The chroma signal is taken from capacitor divider C706-C707 located across the chroma bandpass transformer. The horizontal control pulse is received from the collector circuit of transistor Q701 which is the blanker.

The purpose of the negative-going control pulse is to key the PNP amplifier transistor into conduction during the time color burst is present on its base. The resulting collector information is the amplified 3.59 burst with all other chroma information missing.

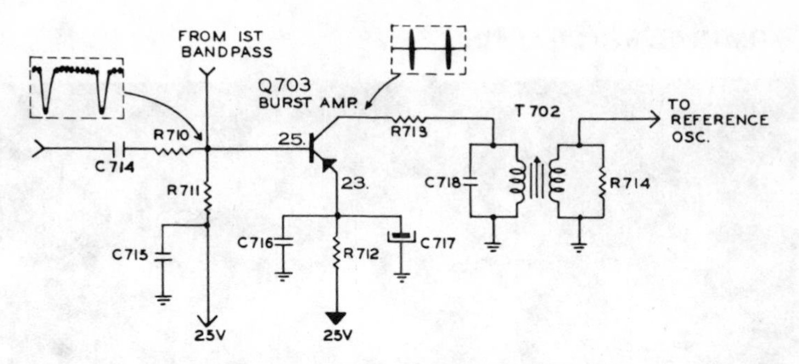

Fig. 9-2. Admiral K20 color-burst amplifier.

The color-burst signal is then coupled to the 3.58 MHz oscillator via burst transformer T702. At this point, the voltage from the burst transformer is used to hold the 3.58 MHz oscillator in step with the incoming burst's phase and frequency. It also supplies bias voltage for the color-killer detector.

Probable Cause: Use a scope to quickly check out the burst circuit. Check for proper keying pulses at the base of Q703 and for correct amplitude of the 3.58 MHz burst information at the collector.

This loss of color was due to an open C714 pulse-coupling capacitor. Other possibilities could be a faulty Q703 burst transistor or a defective T702 burst transformer. After repairs are made, check for zero beat of the 3.58 MHz color reference oscillator and adjust as necessary.

INTERMITTENT COLOR

Picture Symptom: With the AFC switch in the off mode, the picture and color would be good. But when the AFC was switched on, the VHF tuner would be moved off frequency and the color picture would be lost. Look for trouble in the automatic frequency control (AFC) circuit.

Circuit Operation: If the TV station is not tuned in carefully, or if the tuner drifts, a change in phase shift at L403 can occur (Fig. 9-3). For example, if the picture carrier changed to 45.25 MHz, diode D401 would conduct more and D402 would conduct less, resulting in a negative voltage appearing on the AFC line. If the carrier shifted upwards, the diode action would be the opposite, resulting in a positive voltage on the AFC line. This AFC correction voltage is fed to the tuners. Adjust the fine-tuning control with the AFC switch in the off position, then switch it back on for drift-free color reception. This should be done for each VHF channel, since the tuner has preset fine tuning.

The correction voltage produced by the AFC circuit is fed to the VHF tuner, where it is applied to the base of Q5, an NPN

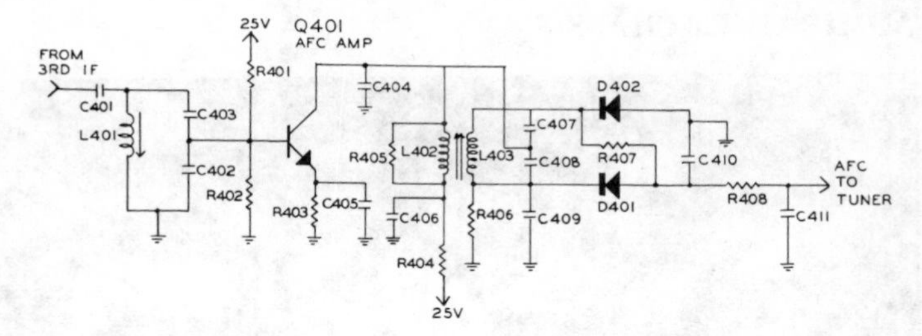

Fig. 9-3. Automatic frequency control circuit.

transistor whose emitter is left disconnected and whose collector is connected to the VHF oscillator tank circuit. Thus the collector-to-base junction serves as the AFC tuning diode. As the AFC voltage varies, the junction acts as a variable capacitor. The correction voltage changes the capacitance of the junction and thus corrects the oscillator error. In the UHF tuner, the AFC diode performs a similar function.

Possible Cause: This tuner frequency shift was caused by a shorted D402 diode in the AFC circuit, which caused an unbalanced condition and fed an incorrect voltage to the tuner.

A defective Q5 transistor (AFC diode) in the tuner could result in no AFC action or could cause the tuner's local oscillator to drift.

ADMIRAL K20 CHASSIS

Picture Symptom: The set produced a good black-and-white picture but no color. Some of the circuits that could cause this loss of color are 3.58 MHz color oscillator, color burst, ACC, chroma bandpass amplifier, and the color-killer stage. After a few checks the problem was isolated to the color killer circuit.

Circuit Operation: The color-killer system (Fig. 9-4) is made up of two stages: color-killer detector Q707 and color-killer amplifier Q708. The killer amplifier acts as a switch to turn on or turn off the second bandpass amplifier. For no color signal, the bias for this transistor is set by the color threshold control.

During a color program, both the killer detector and killer amplifier transistors conduct. As a result, a voltage is produced at the collector of the killer amplifier that provides the forward bias for the second bandpass amplifier. With no burst, this forward bias is not present and the second bandpass amplifier does not conduct—this eliminates any color snow for black-and-white transmissions and between channels.

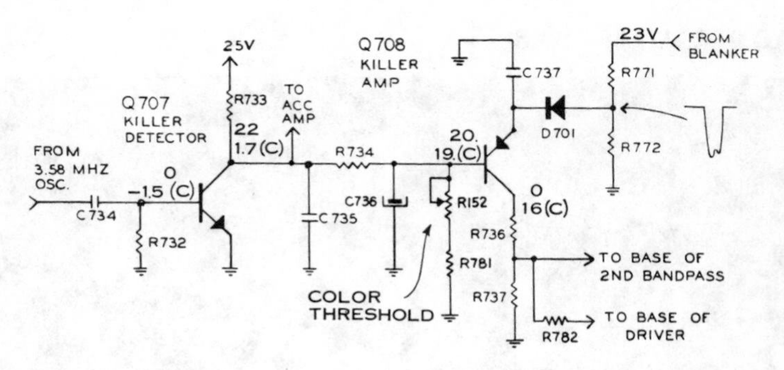

Fig. 9-4. Collor-killer system and burst blanker.

The 3.58 MHz reference oscillator operates continuously, but the output level is increased when the burst signal is present during color programs. Capacitor C734 couples the color oscillator signal to the base of the killer detector Q707. With color, the amplitude of the signal is sufficient to cause base rectification to occur, and the killer detector then conducts, making the collector voltage less positive. This lowers the voltage at the junction of resistors R733 and R734, at the base of the killer amplifier Q708. As the transistor conducts, a positive collector voltage is produced across divider resistors R736 and R737, and the voltage across R737 is applied to the base of the second bandpass amplifier, providing forward bias.

It is necessary to remove the color burst from the chroma signal before the chroma information reaches the demodulators. This is accomplished by shutting off the second bandpass amplifier through the killer amplifier during horizontal retrace time. This pulse reverse-biases the diode, which opens the emitter circuit and stops transistor conduction. Forward bias is removed from the second bandpass amplifier and burst blanking is thus accomplished.

When a black-and-white program is being received, the amplitude of the oscillator signal going to the killer detector is not sufficient to cause conduction. As a result, the killer amplifier is reverse biased and does not produce the turn-on bias for the second bandpass amplifier.

Probable Cause: This no-color symptom was caused by high leakage in capacitor C736, located in base circuit of Q708. A loss of color could also be caused by a defective Q707 killer detector, Q708 killer amplifier transistor, or D701 diode.

Picture Symptom: This set produced too much color and could not be properly regulated with the color level control. Look for trouble in the color level control circuit or automatic chroma control system.

Circuit Operation: The automatic chroma control (ACC) is performed by controlling the gain of first color bandpass transistor Q701 (Fig. 9-5). As the chroma signal from the station is increased, the output of the 3.58 MHz burst signal

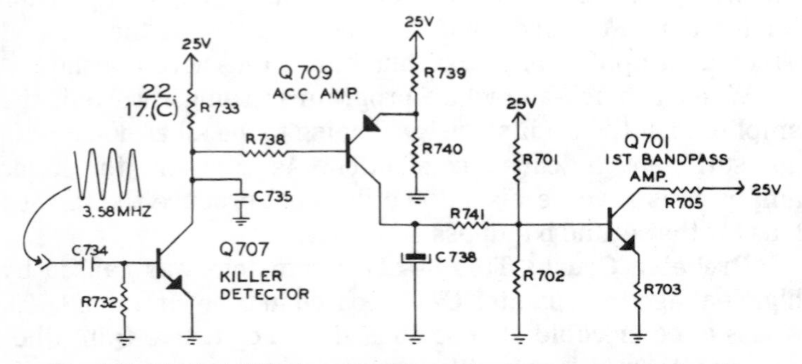

Fig. 9-5. Admiral K20 chassis, automatic chroma control circuit.

from T702 is increased. This increased signal is applied to the base of killer detector Q707, increasing its conduction. The current path is from the collector of Q707 through resistor R733. Because of the filtering action of C735 in the collector circuit, the 3.58 MHz current pulses are averaged, and the voltage developed at the junction of resistors R733 and R738 is now less positive.

ACC amplifier Q709 receives a less positive voltage at its base, which increases its conduction. The increased collector current adds to the current in resistor R702, increasing the voltage to the base of Q701 and reducing its gain.

As the incoming color burst signal decreases, the opposite sequence of events takes place.

Probable Cause: Resistor R701 had increased in value, causing less voltage to appear at the base of Q701. This in turn caused Q701 to conduct more and produced too much color.

Picture Symptom: Loss of all green and most of the red color. A color-bar generator was connected to the set for this check. Look for trouble in the chroma processor module.

Circuit Operation: The chroma processor circuit module (Fig. 9-6) is a complete chroma system in one module and receives its input signal from the emitter-follower output of the video-IF module. The chroma module contains two integrated circuits: IC601 contains a chroma bandpass amplifier, a reference oscillator, an ACC circuit, and a color killer. The other, IC602, is not shown, but contains a chroma demodulator, a color summing matrix, and three output amplifier stages.

The composite video input signal from the IF module is first amplified in a low-gain stage (Q600) having a tuned collector circuit peaked at about 4 MHz. This gives it a rising amplitude characteristic in the chroma passband to compensate for the negative slope of the IF response. An unbypassed emitter resistor gives a stage gain of about 4, with a reasonably high input impedance to reduce the loading on output stage Q204.

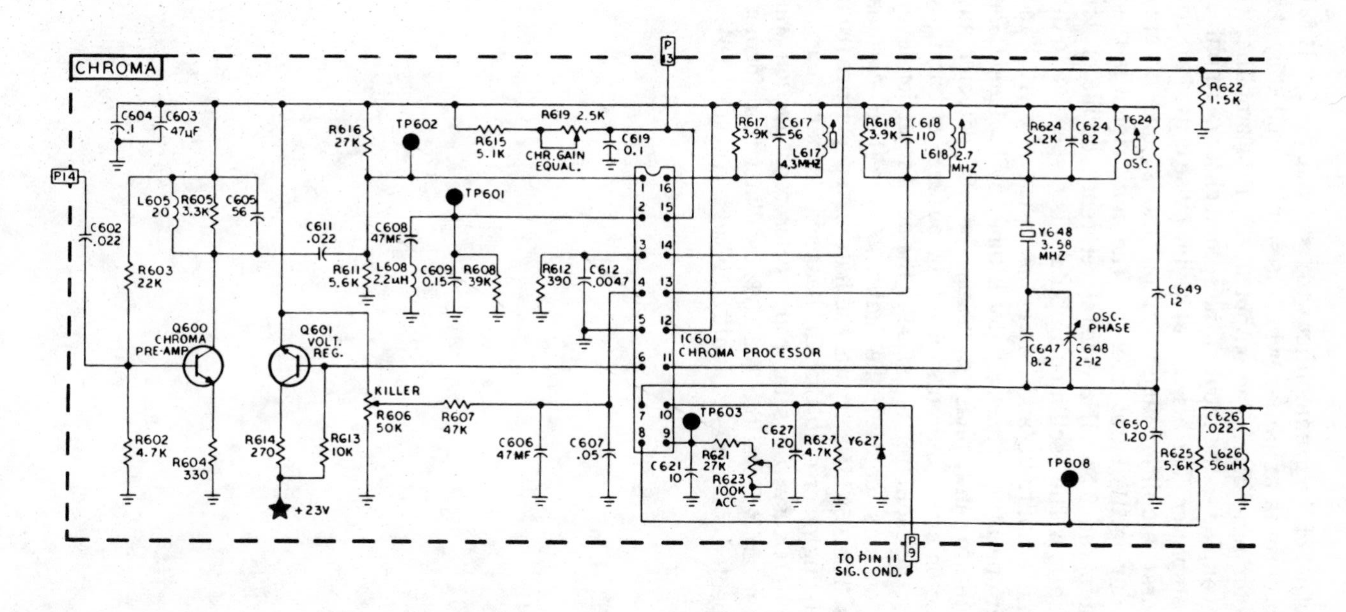

Fig. 9-6. GE MA chassis, chroma circuit.

The output of the preamplifier Q600 goes into pin 1 of IC601 to the first chroma amplifier, which is gain controlled by the ACC detector amplifier. The output of the first amplifier is internally coupled to the input of the second chroma amplifier and burst amplifier. The horizontal keying pulse is used to turn on the burst amplifier and turn off the second chroma amplifier during the burst interval. The burst signal is then injected into the 3.58 MHz crystal reference oscillator for synchronization. The oscillator amplitude is responsive to the burst signal amplitude, and this action is used for ACC and color-killer operation. The oscillator is tuned by T624 and C624, and adjusted to burst phase by C648. The ACC detector and killer detector sense the burst level, or absence of burst, by monitoring the oscillator response to the burst injection level. The two outputs control the gains of the two chroma amplifiers. The 3.58 MHz crystal oscillator is a departure from the crystal-ringing circuits that GE has used for many years.

Probable Cause: This improper color reproduction was caused by a faulty T624, the 3.58 MHz oscillator transformer. Other suspected components would be IC601 and crystal Y648.

Picture Symptom: The picture has too much color and could not be turned down by the color control. This trouble could be on the chroma/video module board or in the color-level control circuits. But this time the problem was located in the new automatic color-averaging circuit.

Circuit Operation: Some color level changes occur when switching channels and are partly corrected in many sets by

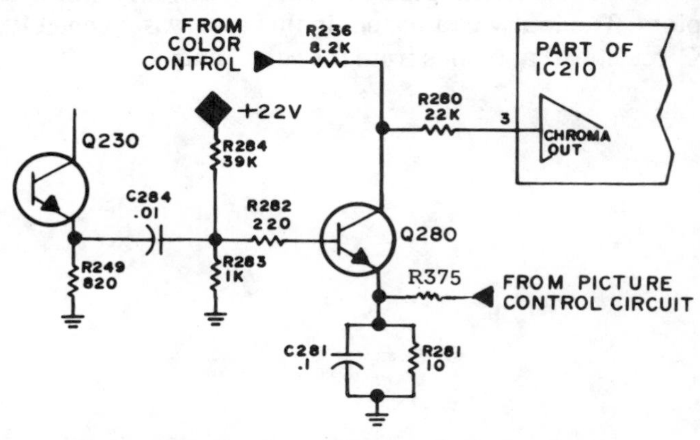

Fig. 9-7. Automatic color averaging circuit (ack-ack).

the automatic color control (ACC) circuits. However, the ACC functions with the burst level and (sad to say) the burst amplitude varies a good bit, so the color level still varies. The GE YA chassis has another automatic correction circuit to help keep the color level constant. This is called the automatic chroma averaging circuit (ACAC)—sometimes referred to as the "ack-ack" circuit.

The ACAC detects the *average* amplitude to the chroma signal and adjusts the chroma gain to obtain a more consistent color level. Note simplified circuit in Fig. 9-7. The base of ACAC transistor Q280 receives the chroma signal from the emitter of color amplifier Q230. The collector of Q280 is connected through R280 to the color control terminal of IC210. By means of R283 and R284, the base of Q280 is biased below the conduction point. The emitter of Q280 is coupled to the picture control circuit by R375, which allows the master picture control circuit to set the threshold point at which Q280 will conduct. When the chroma signal at the base reaches the threshold point, Q280 conducts and its collector voltage decreases, lowering the DC control voltage at pin 3 of IC210 and decreasing the color gain. The circuit time constants are such that instantaneous color-level changes will not affect the gain . Only long-term changes cause the circuit to function, making the average color level more consistent.

Probable Cause: This color trouble was cleared by changing the chroma and video module board. The ACAC circuit is located on this plug-in panel, and some checks were made on the detective board to see what would cause this symptom. The faulty component in this case was an open R236 (8.2K) resistor that comes from the color control.

Picture Symptom: Some chroma phase shift. Use the following troubleshooting procedures to isolate this color fault.

Circuit Operation: For no color or insufficient color, make both the ACC and APC adjustments. If these adjustments do not react correctly, the subcarrier regenerator (SCRC) module (Fig. 9-8) is at fault when the problem is no color.

Should both the ACC and APC adjustments react properly, the trouble is then likely to be insufficient color. Note: It is possible that insufficient color is accompanied by poor color sync, but it does not matter which symptom is used in this procedure because the problem area will be found in any event. To isolate, check the ACC voltage between TP6 and TP9 with the set correctly fine tuned to a color program. If this voltage is greater than 0.1V, the SCRC module is at fault. If the voltage is between 0.05 and 0.1 volt, however, the chroma-IF (CIFC) module is defective.

If the ACC adjustment can be completed but the APC adjustment cannot, the problem must be no color. The fault may be the chroma-IF module, chroma demodulator/blanker

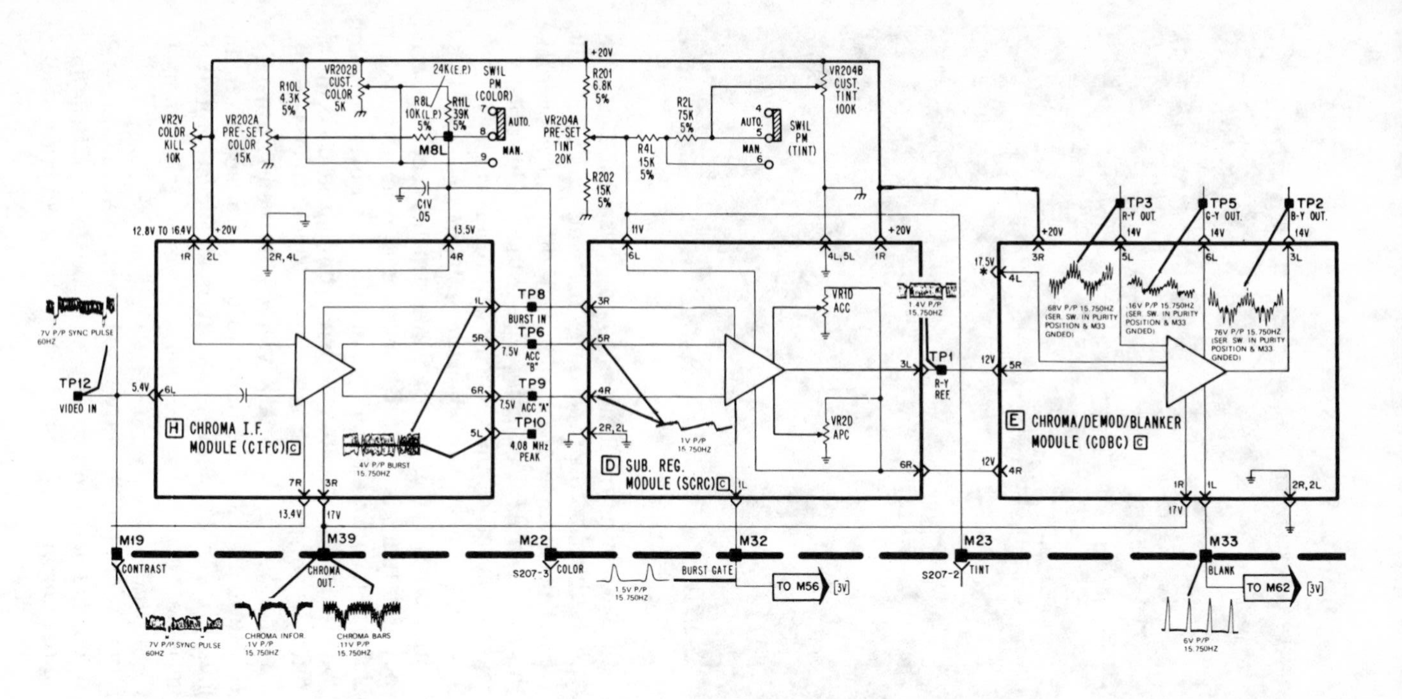

Fig. 9-8. Philco/Ford 3CY91, block color circuits.

(CDBC) module, or main chassis circuitry. To isolate, check the chroma signal output from the chroma-IF module to the demodulator module at terminal M39. It must be about 0.3V P-P on the scope. If not, the trouble is the chroma-IF module or circuitry from the low-level video (LLVC) module to the chroma-IF module. To isolate, check the signal input to the chroma-IF module at TP12. The signal input at TP12 must be about 3V P-P for the composite video. If it is not, the trouble is ahead of the chroma-IF module, in the main chassis board connections, assuming that there is no fault in the low-level video module.

If the ACC adjustment produces the correct action but the APC adjustment does not, the trouble will again be insufficient color. The weak color symptoms may also be accompanied by poor color sync. In either case, the trouble must be in the 3.58 MHz subcarrier regenerator module.

For color sync troubles, make the ACC and APC adjustments first. If both the ACC and APC adjustments can be completed, the trouble must involve the chroma IF-signal from terminal 1L on chroma-IF module to terminal 3R on 3.58 MHz regenerator module. Use an oscilloscope to pinpoint the exact location of the trouble. If the ACC adjustment can be completed but the PAC adjustment cannot be, the trouble is in the 3.58 MHz regenerator module. But if the ACC adjustment will not produce the correct results but the APC adjustment does, there is almost certainly insufficient color, which may also be accompanied by poor color sync. This would indicate a faulty regenerator module.

For color tint troubles, if the tint control does not function or will not obtain proper flesh tones, check for a 6-to-12 volt DC variation between terminal M23 and ground as the tint control is turned through its range. Turn the Philcomatic control off while this is done. If the voltage does not read properly, the trouble is in the tint control. In these chassis, tint problems will not usually be found if the tint control circuit is good. It may, however, be possible to get some chroma phase shift as the result of trouble in the chroma IF or the demodulator module. This, of course, results in color tints that are improper. Usually, defective demodulators result in improper white balance so the troubleshooting procedure for isolating brightness and white balance trouble is used. CRT bias checks should also be made for these troubles.

If the color killer does not work properly, check the DC voltage at 1R on the chroma-IF module. The voltage should vary between about 12 to 20 volts DC. If it does not vary, the trouble is in the color killer control VR2V or associated circuitry. If it does vary, the chroma IF module is at fault.

Probable Cause: This color phase shift trouble was caused by a defective subcarrier regenerator module (SCRC).

Picture Symptom: Color picture was weak and not very clear, with fine horizontal streaks going across the screen. This symptom was of an intermittent nature. Look for trouble in the color processing circuits.

Circuit Operation: Let's now run the chroma signal through the color circuit path by referring to the schematic in Fig. 9-9.

Video from pin 11 of IC301 is fed to the primary of color-IF transformer T601 through capacitor C601. The secondary winding is capacitor coupled to color processor IC601 at pin 15. One end of each winding is returned to ground. Sync at pin 16 is divided by resistors and fed into the IC at pin 2 for application to the pulse former stage. The resulting output pulses at pin 3 are timed and shaped by components on the board before being fed to point 13 of IC601.

Color processor IC601 performs all functions of color processing, from color-IF in to color difference signals out, with relatively few external components.

The color-IF signal is amplified and applied to a gain-controlled amplifier within the IC. Gain is determined by

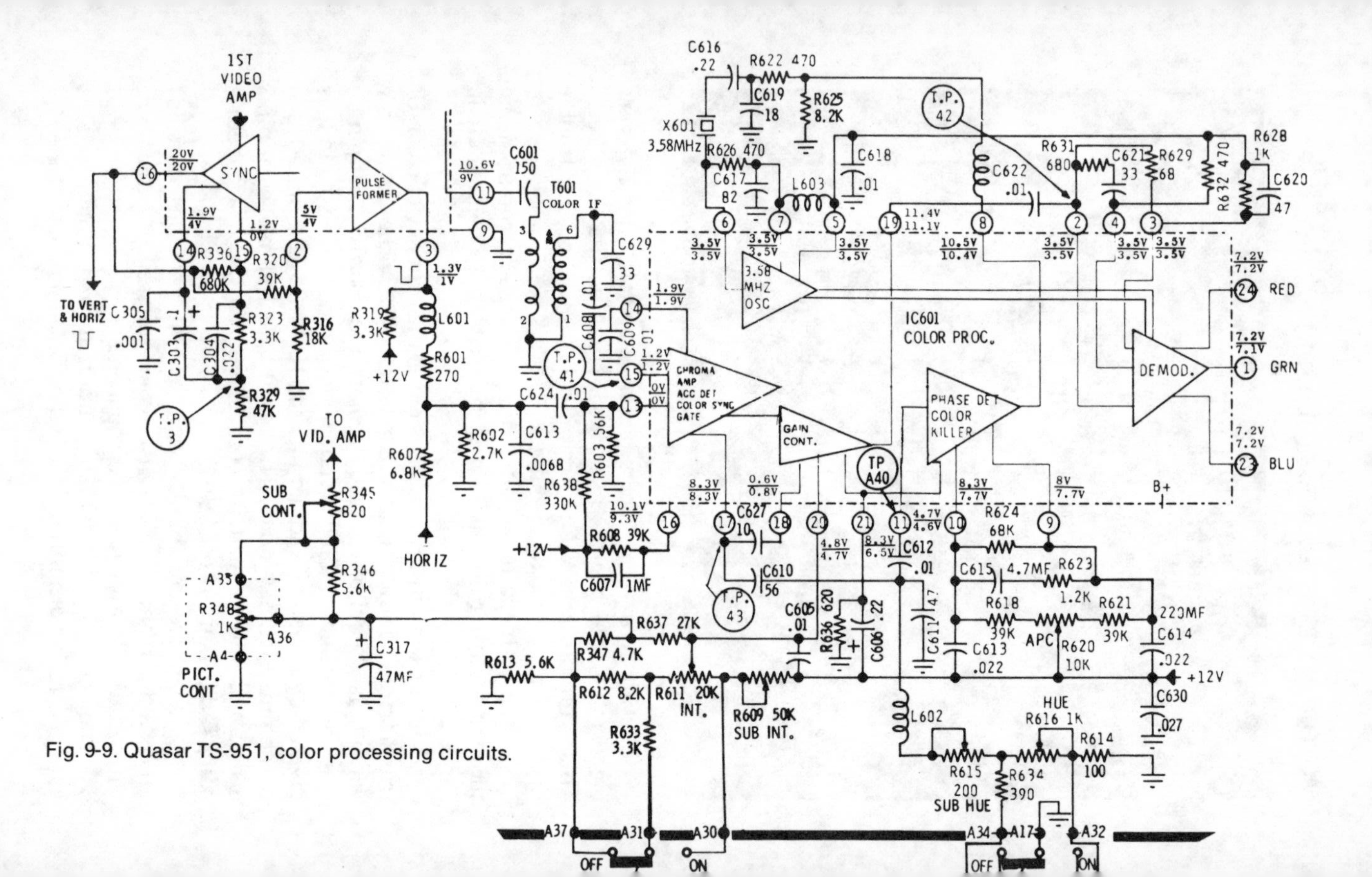

Fig. 9-9. Quasar TS-951, color processing circuits.

the DC voltage applied to point 20 from the intensity control. The color sync signal is separated (gated) from the color-IF signal and leaves the IC at pin 17, where it is capacitively coupled back in at pin 11, being fed to the phase detector and color killer section. When color sync is present, the chroma control stage operates, and in the absence of color signals, the killer turns the control stage off.

The hue is varied by phase-shifting the color sync signal fed to pin 11. A variable control in the APC provides a means of balancing the DC voltages to the phase detector via pins 9 and 10.

The color sync signal, controlled by ACC and PAC, leaves the IC at pin 8 to ring a crystal. The resulting CW signal from the crystal enters the IC at pin 6 to control the 3.58 MHz oscillator. Two different phases of this CW signal output are fed to the color demodulators.

The color-IF signal from the gain controlled section leaves the IC at pin 19. It is coupled back into the IC at pins 2, 3, and 4 for presentation to the color demodulators.

Color-difference signals are recovered in the demodulator by demodulating the phase and amplitude of the color-IF information with the correct phase of 3.58 MHz CW from the oscillator. The resultant color-difference signals leave the IC at pins 24 (red), 1 (green), and 23 (blue).

Probable Cause: This streaked, low color level picture was due to leakage in C601, a 150 pF coupling capacitor from point 11 of IC301 to pin 3 of color-IF transformer T601.

Picture Symptom: Too much blue and focus is poor. Look for trouble in the color demodulators and amplifier stages. Also check the condition of the color picture tube.

Circuit Operation: Three diode demodulators (Fig. 9-10) are used to demodulate the chroma signal. The demodulator color-difference signals are applied to individual driver and output stages, which in turn are capacitor coupled to the CRT control grids. The DC level is restored to the color signal by diode clamp action.

Fed to each demodulator circuit are two signal voltages. One is a 3.58 MHz CW reference signal, the other a chroma signal from the bandpass amplifier. The phase of the reference 3.58 MHz signal is shifted a specific amount with respect to burst for each demodulator circuit. This lets each demodulator extract the appropriate color-difference signal from the applied chroma information.

The driver stage uses an emitter follower to provide a proper match between the relatively high output impedance of the demodulator circuit and the low input impedance of the output stage. The 3.58 MHz ripple component is attenuated

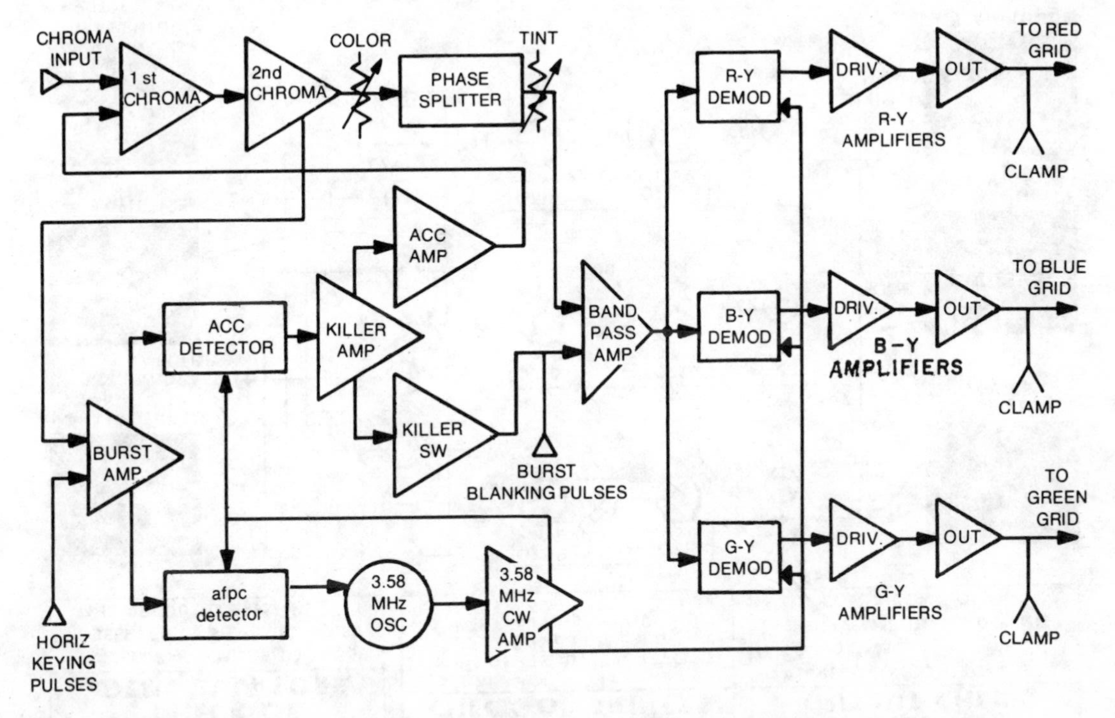

Fig. 9-10. CTC-40, RCA chroma block diagram.

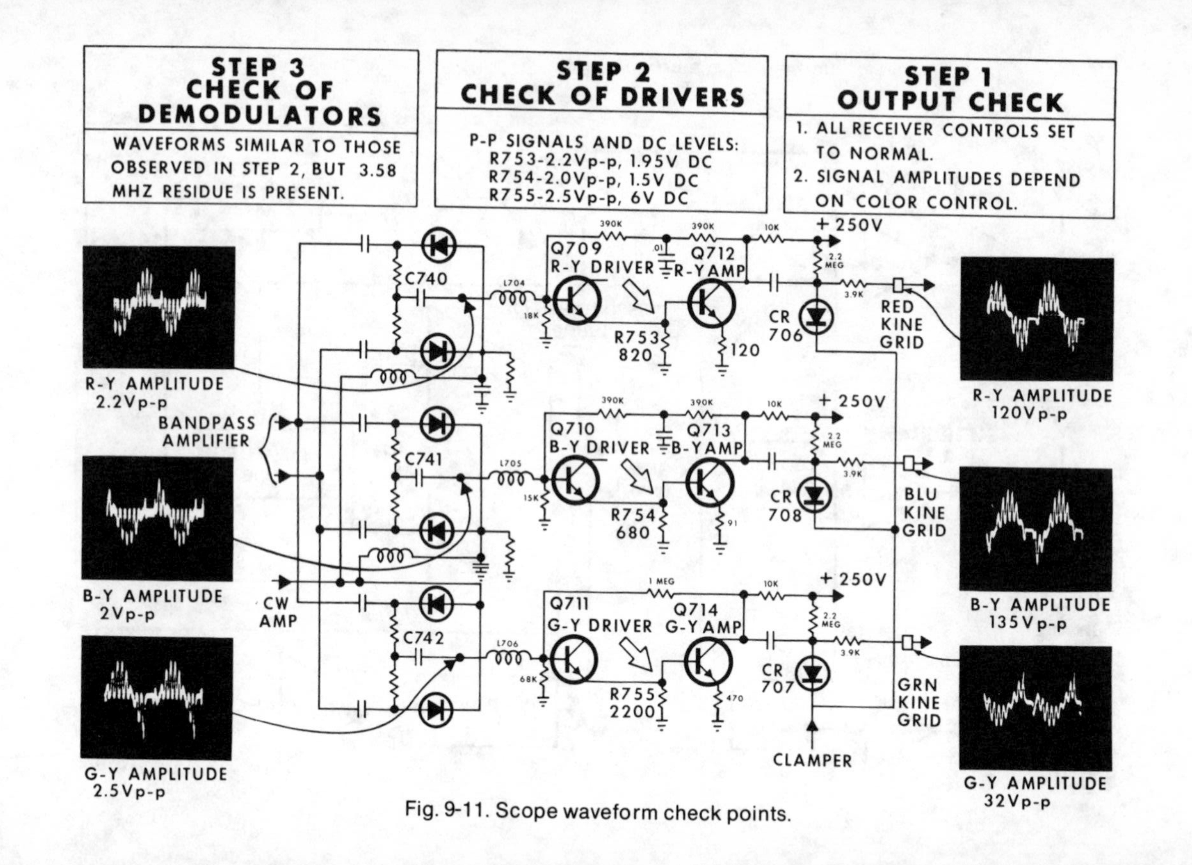

Fig. 9-11. Scope waveform check points.

from the output of the demodulator circuit by a low-pass filter consisting of L704 and the input impedance of the driver stage. Base bias for the driver stage is provided by a voltage divider network.

The demodulated output signal is coupled via a capacitor and a 3.9K resistor to the control grid of the CRT. Note that the clamper diode is also located in this circuit.

Color Demodulator and Amplifier Checks: Refer to the waveforms and demodulator schematic in Fig. 9-11. Use a scope to check the three output signals from the chroma board. The amplitudes of these outputs will depend on the setting of the color control. The null points of the three outputs may be varied over a range of at least 75 degrees. The color control was set to produce 120 volts of R-Y output for these scope waveforms.

The waveforms observed in step 2 of Fig. 9-11 are similar to the ones in step 1, except they are inverted and their amplitudes are much lower and nearly equal. Changes in the DC voltage would probably be due to circuit failure and would cause a change in the gains of the output amplifiers, shifting the color balance.

The outputs of the three demodulators are observed in step 3. These waveforms differ from the ones in step 2 in that the 3.58 MHz ripple is present. Failure in a demodulator could cause partial or complete loss in amplitude of a signal, or a change in the null point. These color faults can be demonstrated by jumpering a demodulator diode, a diode load resistor, or a phase shifter coil.

Probable Cause: This blue picture symptom was caused by a defective (leaky) B-Y transistor amplifier Q713. A shorted clamp diode, such as CR708, may also cause the same problem.

Picture Symptom: No color information, but good black-and-white picture and sound. Look for trouble in the chroma circuits.

Circuit Operation: Refer to the color block diagram in Fig. 9-12 for paths of chroma information after detector C1. Video information goes through the Y-amplifier circuitry to drive the cathodes of the CRT, while chroma information flows through the 5 pF capacitor to the first color amplifier, then via the second and third color amplifiers, demodulators, and R-Y, B-Y, and G-Y amplifiers to drive the grids of the CRT.

From the plate of the first color amplifier, burst and color signals are directed toward the 6JC6 burst-amplifier grid, where the burst information is separated by a keying pulse from the horizontal circuitry. Burst is mixed with 3.58 MHz signal in phase-detector diodes of the ACC killer to create a DC color-killer voltage at test point Q, the grid of the first color amplifier. Also, the burst is mixed with the 3.58 MHz signal in diodes Z12 and Z13 of the AFC phase detector to create a DC color-locking voltage at test point W to keep the 3.58 MHz oscillator locked on frequency and in phase.

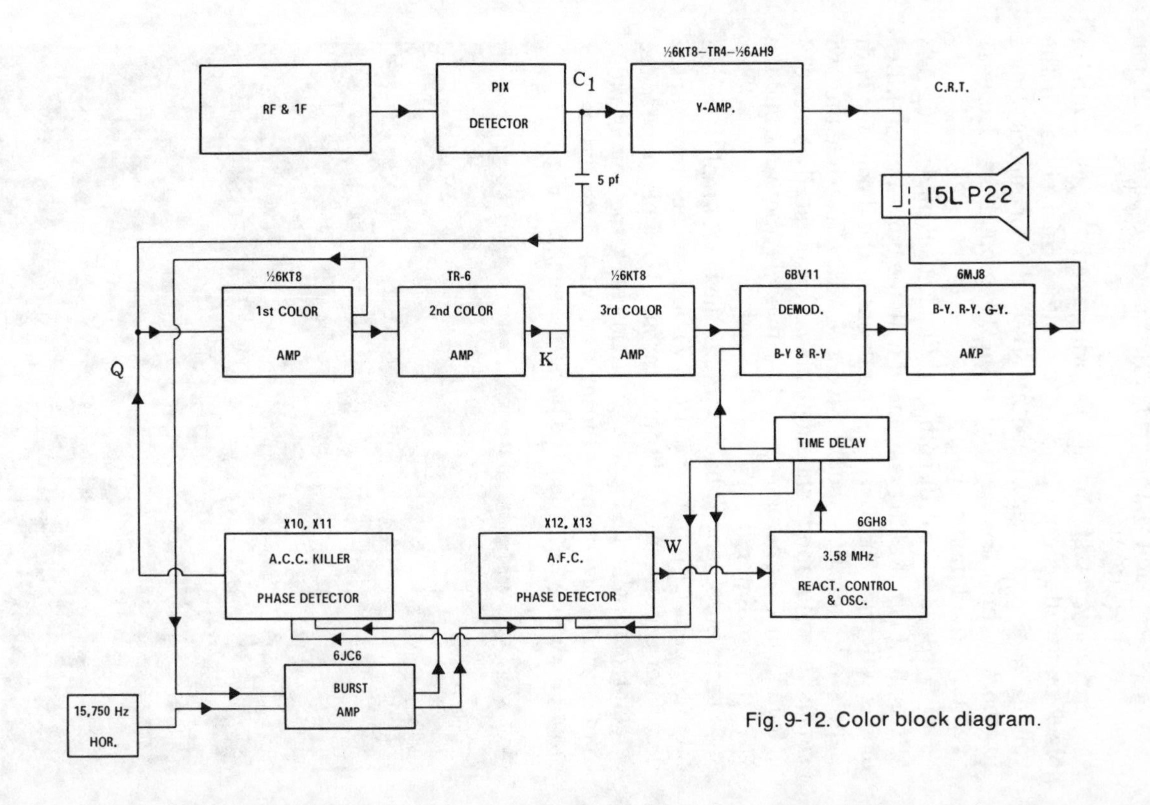

Fig. 9-12. Color block diagram.

Service Checks and Tips: Use a color-bar generator to feed an RF signal into the tuner, switching it from a color to a no-color signal as your make you measurements. Connect a VTVM at test point Q, the grid of first color amplifier V1B, pin 7.

Normally, test point Q will read –0.6V with a black-and-white signal, and –6.0V with a color signal. If the voltage at Q reads zero in both B&W and color modes, check for shorted capacitor C89 from point Q to ground, or a decrease in value of the 1 meg resistor.

If Q reads –0.6V in both B&W and color modes, the 3.58 MHz signal is not being compared with the burst in ACC killer diodes X10 and X11. Use a scope to check the diodes for both signals. The junction should have 3.58 MHz CW, while the two ends in the color mode should display color burst. If no 3.58 CW is found, then scope back into the time-delay network and the 3.58 MHz oscillator.

For no burst, check back through the burst amplifier plate coil into the 6JC6 burst amplifier (V14), then back through the first color amplifier for chroma, through the 5 pF to the video detector output at test point C1. Also check the keying pulse from horizontal sweep to the grid of V14. Note the correct timing of the keying pulse and burst shown in the Fig. 9-13.

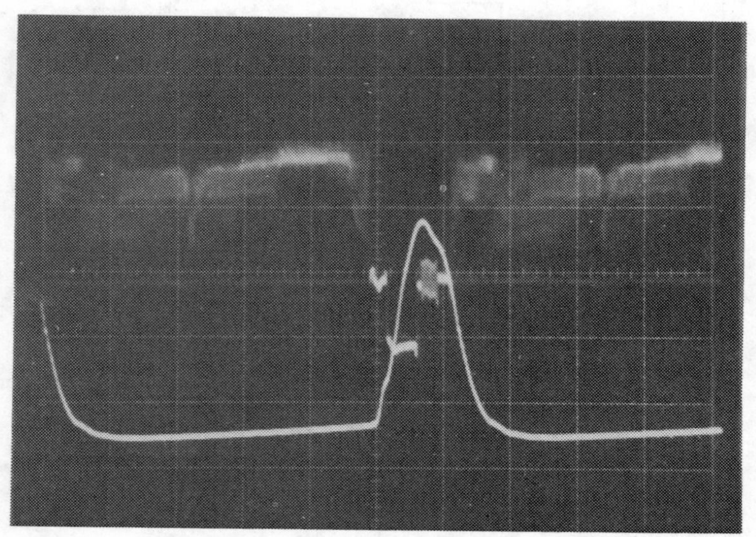

Fig. 9-13. Dual-trace scope trace showing burst and keying pulse with correct timing.

If the scope shows proper amplitudes for both the 3.58 MHz and the burst signal on the ACC killer phase-detector diodes, the problem may be caused by an out-of-phase condition. To check, turn the color killer full on, then note if color bars are displayed in the color mode, indicating the 3.58 MHz oscillator is not synchronized. If this is the case, you may only need to zero-beat the color oscillator by adjusting coil L47 at pin 1 of reactance tube V15A (6GH8A).

When test point Q reads −0.6V on B&W, and −5 to −7 volts in the color mode, then the burst is being compared with the 3.58 MHz signal in the ACC killer diodes. This indicates the 3.58 MHz oscillator is in good shape, as is the burst amplifier, the ACC killer diodes, and the first color amplifier.

If the color amplifier circuitry is being gated into the color mode, but is not functioning as such, then make sure the color killer setting is high enough. Check the second color amplifier (TR6), third color amplifier (V6B), demodulator sections (6BV11), and the B-Y, R-Y, and G-Y amplifiers (6MJ8). Also check the B+ and bias voltages. Use your scope to trace the color-bar generator signal through the chroma stages.

In some cases of no color, the chroma circuitry is not gated into the color mode. In this case the voltage at test point Q does not change, and the problem must be traced back through the ACC killer diodes circuit.

Probable Cause: This loss of color was caused by an open 5.6K resistor in the screen-grid circuit of burst amplifier tube V14. This caused a loss of 3.58 MHz burst signal to L45, and AFC phase-detector transformer. Check or replace the 6JC6 tube, as it may have shorted and caused the 5.6K resistor to fail.

ZENITH 19FC45 CHASSIS

Picture Symptom: Picture is usually pulled down from the top (may also be pulled up from bottom) and has lots of black horizontal streaks going across the screen. This is usually an intermittent trouble and will generally show up after the set has warmed up about 10 or 15 minutes. Look for trouble in vertical sweep module 9-90.

Circuit Operation: The oscillator stage in Fig. 9-14 consists of transistor Q701 and Q702. When the set is turned on, Q701 begins to conduct, which allows Q702 to draw base current and also conduct. With transistor Q702 conducting, capacitor C704 begins to charge, thus increasing the positive potential on the base of transistor Q701. Since Q701 is a PNP transistor, it will cut off when its base voltage becomes sufficiently positive, then transistor Q702 will no longer draw base current and will also cut off. Capacitor C704 will then discharge through the vertical hold control until the base voltage of Q701 drops low enough to cut if off, allowing the cycle to repeat.

Transistors Q704 and Q705 are connected as a differential amplifier. The positive-going sawtooth waveform at the collector of Q703 is capacitor coupled to the base of Q704, while sample of the yoke current is fed back to the base of Q705. Now the feedback waveform from the yoke should be identical to the oscillator sawtooth. Any error in the feedback waveform will be sensed by the amplifier and will cause Q704 to increase or decrease conduction, thus correcting and maintaining linearity.

The collector of Q704 is coupled to the base of driver transistor Q706, producing a negative-going sawtooth waveform. Q706 them amplifies and inverts the negative-going sawtooth waveform, coupling it to the base of output transistors Q707 and Q708.

During the first half of the sawtooth waveform, transistor Q707 conducts, providing vertical deflection. During the second half of the sawtooth waveform, Q708 conducts, providing vertical deflection for the bottom half of the raster.

Probable Cause: This trouble was caused by an increase in value of R730. The dark lines across the screen were due to an interval arc in R730. A leaky Q707 transistor probably caused this 4.7Ω emitter resistor to overheat. Both output transistors and both emitter resistors should be changed to complete the repair. The same picture symptoms could be caused by faulty set-up switch, an arc in the yoke winding, or a defective pincushion coil.

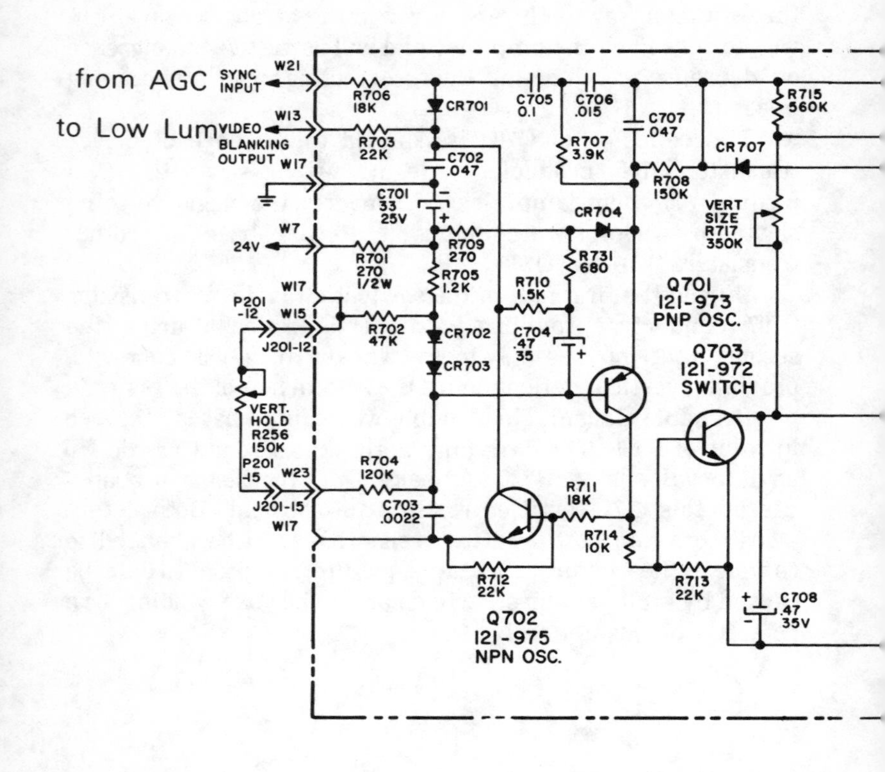

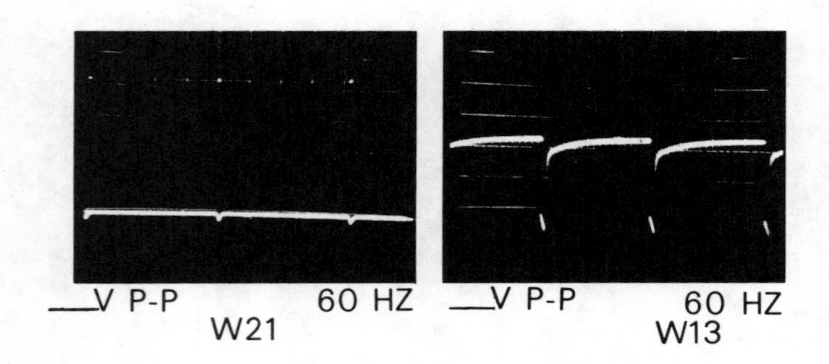

___V P-P 60 HZ
W21

___V P-P 60 HZ
W13

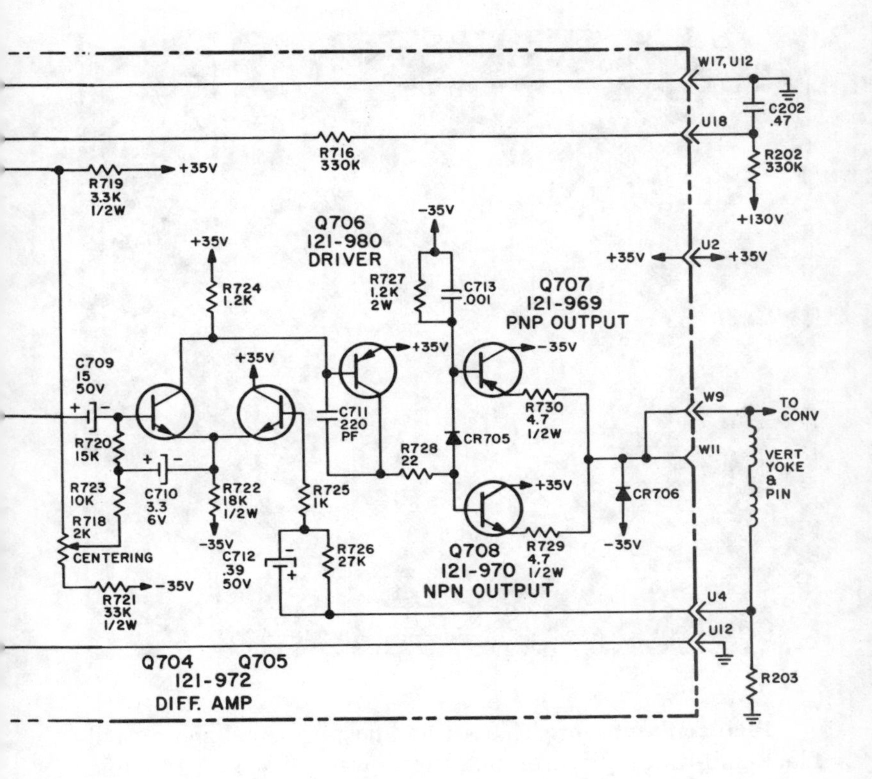

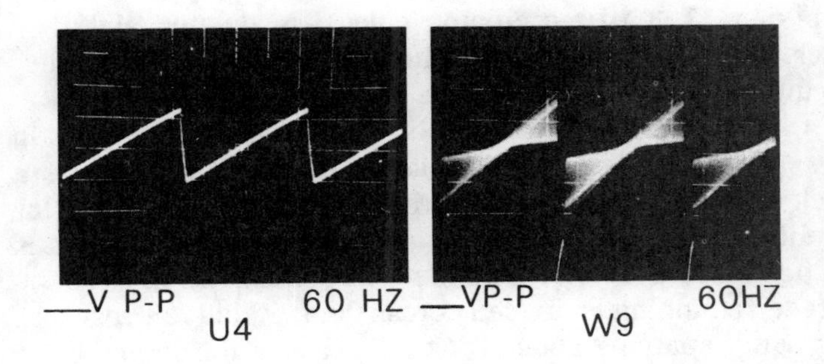

Fig. 9-14. Zenith 9-90 vertical module circuit.

ZENITH 14Z8C50 CHASSIS

Picture Symptom: This set had normal sound and a good black-and-white picture, but there was no color. Possible circuits that could cause loss of color would be color amplifiers, 3.58 MHz oscillator, color demodulator, and color killer. This loss of color was due to a fault in the color-killer circuit, so let's look at its action.

Circuit Operation: Refer to the chroma amplifier circuit in Fig. 9-15 as we follow the color-killer operation. During a black-and-white picture the screen voltage at the first color amplifier is low (near 70 volts) and is reflected to test point K through an 82K resistor, and the color-killer control. Here the voltage remaining is divided across the 4.7K and 22K resistors to ground, applying about 11V to the base of the second color amplifier. The 24V supply is applied to the collector of the second color amplifier through the collector transformer and is also divided across two 2.2K resistors, applying 12.4V to the emitter. Thus, since the base voltage is less positive than the emitter, TR8 (an NPN transistor) is cutoff. Diode X11 will not

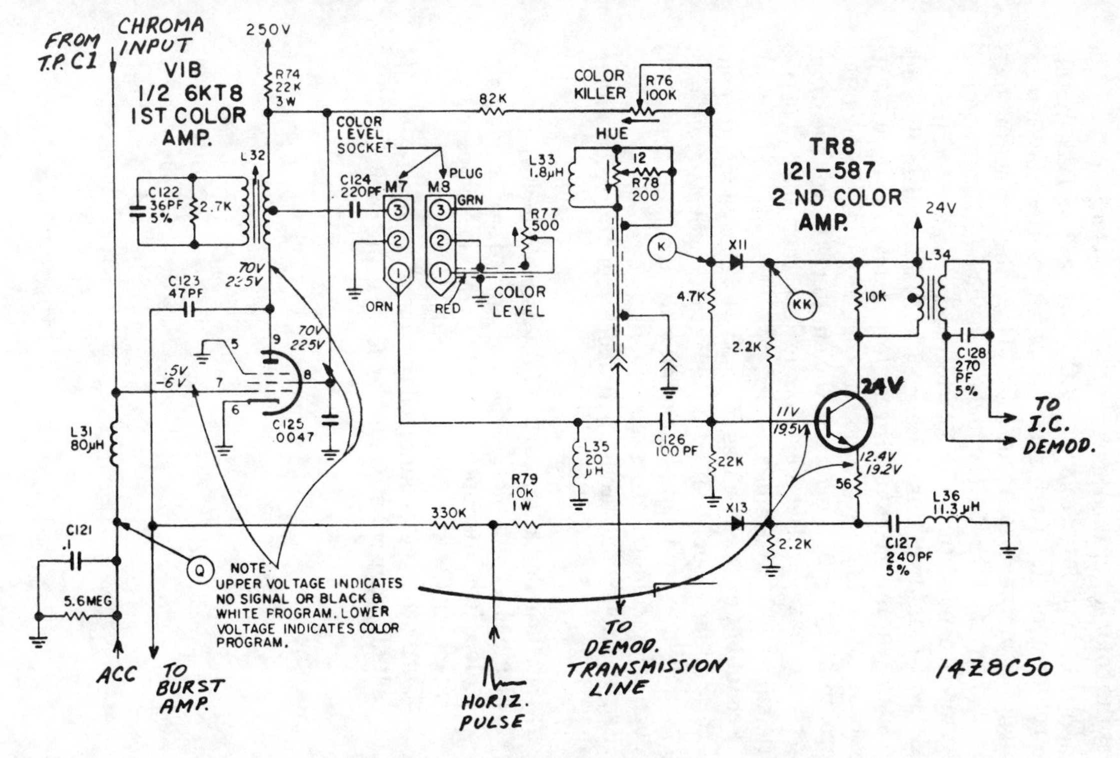

Fig. 9-15. Zenith 15Z8C50, chroma amplifier circuit.

conduct during a monochrome signal because the voltage produced at test point K never exceeds the voltage at test point KK to overcome the 24V on the diodes cathode. X11 is referred to as the color-killer diode.

During a color program, ACC voltage applied to the control grid of the first color amplifier causes the screen voltage to increase to about 225V. Thus, the voltage at test point K increases above 24V and causes diode X11 to conduct. All that is needed to cause X11 to conduct is 24.5V, and the diode will clamp at this voltage. The 24.5V is then divided across the same 4.7K and 22K resistors and results in 19.5V at the base of TR8. By virtue of conduction through TR8, approximately 19.2V appears on the emitter. The voltage drop across the base-emitter junction is less than 0.5V, but is sufficient for TR8 to conduct.

Probable Cause: The loss of color was caused by an open X11 color-killer diode. If diode X11 shorts, the color amplifiers will be active even during black-and-white transmissions and produce colored snow across the screen.

Use the following steps to properly set the color killer level:

1—Set the color-killer control fully counterclockwise.

2—Tune in a color picture. Detune toward smear, until all color just disappears and the picture appears in black-and-white only.

3—Turn the color-killer adjustment clockwise until satisfactory color content appears in the picture.

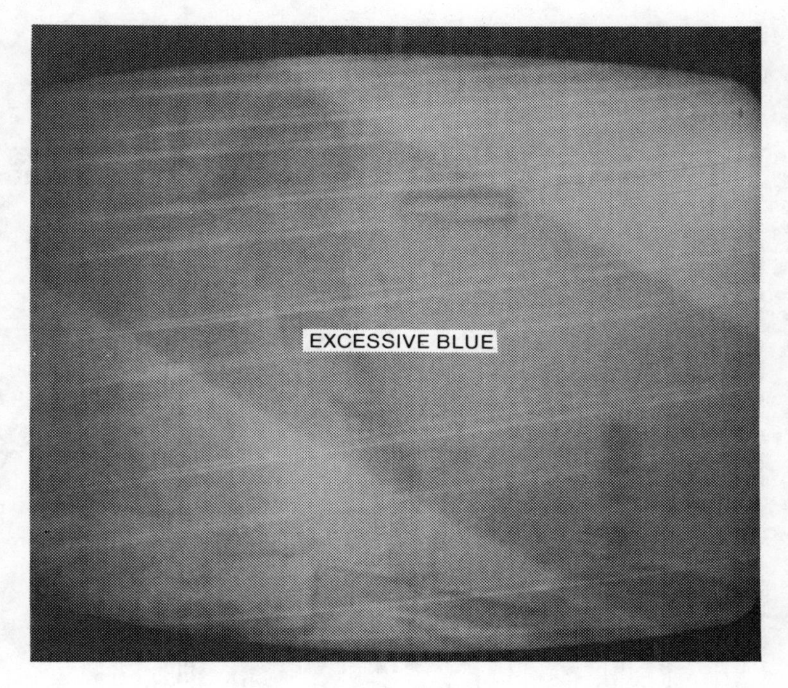

Picture Symptom: Picture very blue, blurred, and with some retrace lines. Look for trouble in the blue amplifier stage located in video output module 9-121. This module is on the picture tube socket.

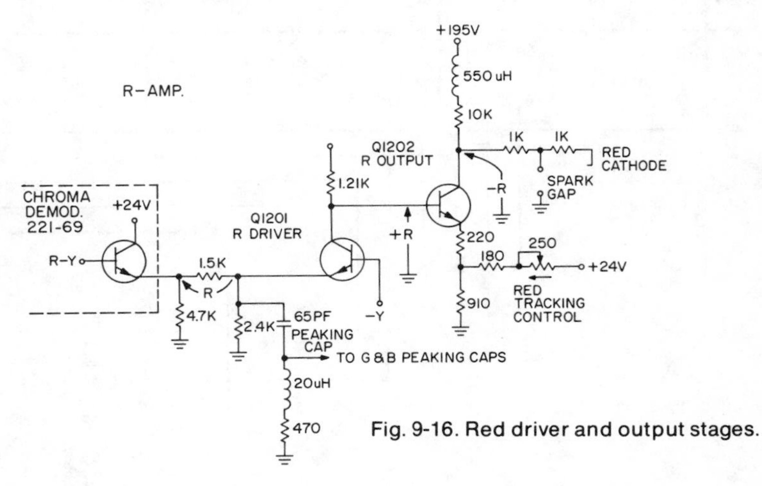

Fig. 9-16. Red driver and output stages.

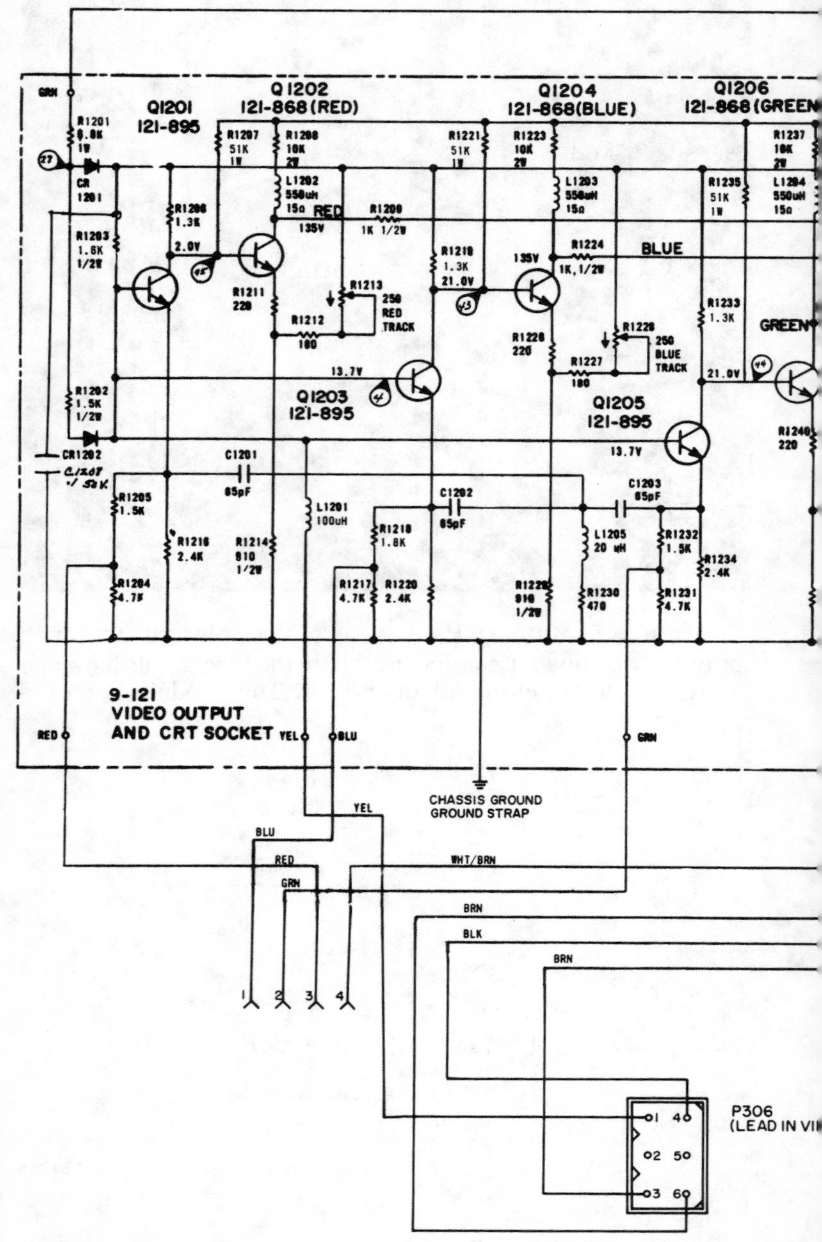

Q1201
121-895
Q1202
121-868 (RED)
Q1203
121-895
Q1204
121-868 (BLUE)
Q1205
121-895
Q1206
121-868 (GREEN
RED
BLUE
GREEN
250
RED
TRACK
250
BLUE
TRACK
9-121
VIDEO OUTPUT
AND CRT SOCKET
CHASSIS GROUND
GROUND STRAP
P306
(LEAD IN VI

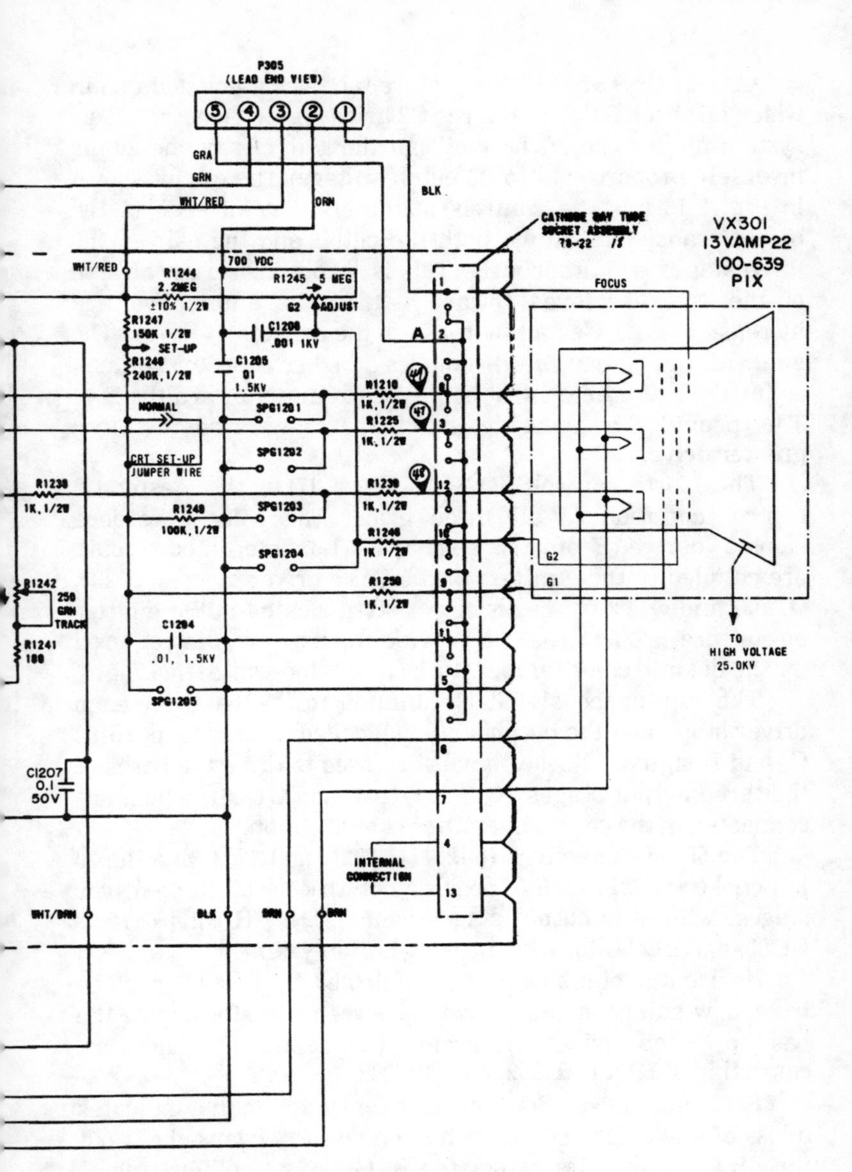

Fig. 9-17. Video output module 9-121.

Circuit Operation: This picture tube has a unitized gun in which all three G1 and all three G2 grids are tied together. This system makes use of the fact that the gain of any one gun is inversely proportional to its cutoff voltage. The circuit shown in Fig. 9-16 has the controls in the emitters of each of the output transistors adjust both the cutoff and the gain of the three stages simultaneously. This is accomplished as the arm of the control moves toward +24V. The emitter voltage increase also causes an increase in the collector voltage. The gun with the higher cutoff requires a higher collector voltage, so for that gun the control is positioned more toward the B+. This provides a higher gain for that stage and therefore greater drive.

The fourth control (R1245 in Fig. 9-17) is the master G2 control common to all three grids. The color difference signals, derived from the standard 221-69 integrated circuit, are coupled to the emitters of the three driver stages (Q1201, Q1203, and Q1205). The three resistors located in the emitter circuit are proportioned to provide the normal gain required for the desired color temperature (black-and-white tracking).

The luminance signal is common to the bases of each driver stage and the outputs at each collector are standard R, G, and B signals. These signals are used to drive the bases of the three output stages (Q1201, Q1204, and Q1206), which are connected in the common-emitter configuration.

The 65 pF capacitors (C1201, C1202, and C1203) and the 20 μH choke (L1201) form a peaking network for all three driver stages, while the output stages contain 550 μH coils (L1202, L1203, and L1204) for total peaking of the system.

Horizontal blanking is accomplished by feeding a pulse from a winding on the horizontal sweep transformer to the base of each driver transistor by means of a network consisting of R1201, R1202, and CR1202.

Probable Cause: For any color set symptom that indicates a loss of one color or too much of one color, always check out the CRT and then its associated R, G, or B amplifier stages. Bias voltage checks on the CRT grids and cathode are very helpful in locating this type symptom.

This blue and blurry picture was caused by a shorted CR1202 horizontal blanking diode and a faulty Q1203 blue driver transistor.

Picture Symptom: A wide green streak across the screen at high brightness levels. May be intermittent in nature or include red or blue streaks. Other symptoms could be too much or loss of color. Look for trouble in video output module 9-89.

Circuit Operation: The color video output module (Fig. 9-18) is fed a composite video signal at terminal W1 and three color difference signals at terminals W15, W19, and W23. The module combines each of the color difference signals individually with the composite video signal. Thus, the module has three output signals that are coupled to the cathodes of the CRT.

Because the same video (luminance) information is fed to the emitter of each color video output transistor, each of the transistors has the same video output signal in the absence of a color signal. Thus, each of the three CRT guns will conduct at the same level and a black-and-white picture will be produced. A change in video information causes all three color video output transistors to change the same amount, which in turn causes the output of the three CRT guns to change the same amount and also sets the gray scale tracking correctly.

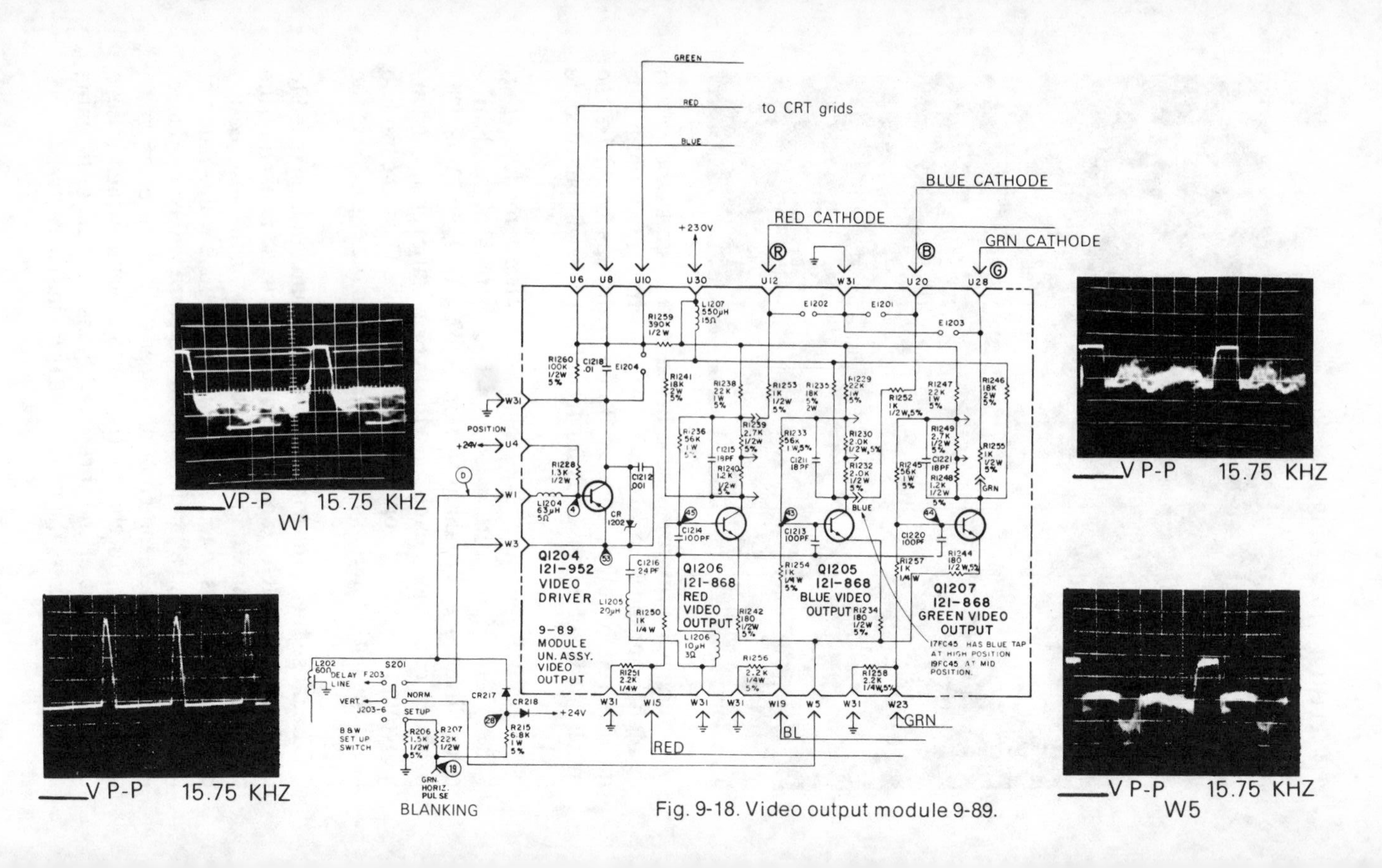

Fig. 9-18. Video output module 9-89.

Chroma information is fed to the base of each color video output transistor from the chroma demodulator IC. Unlike the video information, each color video output transistor receives a different chroma signal, producing the desired color picture.

Other defects that can result within this stage are weak colors, loss of colors, blooming, and excess of one or more colors. For weak or loss of colors that produce a tinted raster, one or more of the RGB video output transistors could be defective—check by substituting a good transistor. But if a good transistor is not available, check by switching the position of the output transistors in question with one of the other transistors. If the absent color is now present and one of the other colors is lost, then the transistor is faulty. However, if the same color is still missing, then check the base and collector circuits for defective components.

For excessive color, isolate the color video output circuitry by the steps that follow:

- Place the set-up switch in the set-up position.
- Turn the G2 controls to minimum.
- Should the corresponding color line of the affected video output stage not extinguish, the video output transistor may be shorted. For a quick check you can switch the RGB video output transistors around.

If the transistors are not defective, check the B+ circuit of each collector. Each of the three video output transistors has two B+ paths to the collector. If either path opens the collector voltage will decrease by 30 to 60 volts, causing one CRT gun to draw too much current and tinting the raster.

Probable Cause: The green streak in this picture was caused by a defective Q1207 green video output transistor. If the streak is red, check Q1206, and for blue, Q1205. If transistor Q1207 is good, check value changes of resistors R1244, R1257, and R1245.

Picture Symptom: This set produced a good black-and-white picture, but there was a complete loss of color. Look for trouble in the color killer, 3.58 MHz oscillator, color amplifiers, or color demodulators. This loss of color was located in the color amplifier (bandpass) section.

Circuit Operation: This chassis utilizes the hybrid color amplifier system shown in Fig. 9-19. The chroma signal is coupled to the grid of the first color amplifier (V1B) via 12 and 5 pF capacitors and an LR network from the 4.5 MHz trap. Shunt peaking is used in the input grid circuit to reduce cross talk and provides better centering of L25, the first color amplifier coil—this helps to provide equal sideband amplification and thus enhances the phase response. Test point Q (grid of first color amplifier) is returned through a 1-meg resistor to ground, to complete the ground return of the control grid.

The color signal is amplified by V1B, and the output of V1B is coupled to the burst amplifier, for burst separation, and to

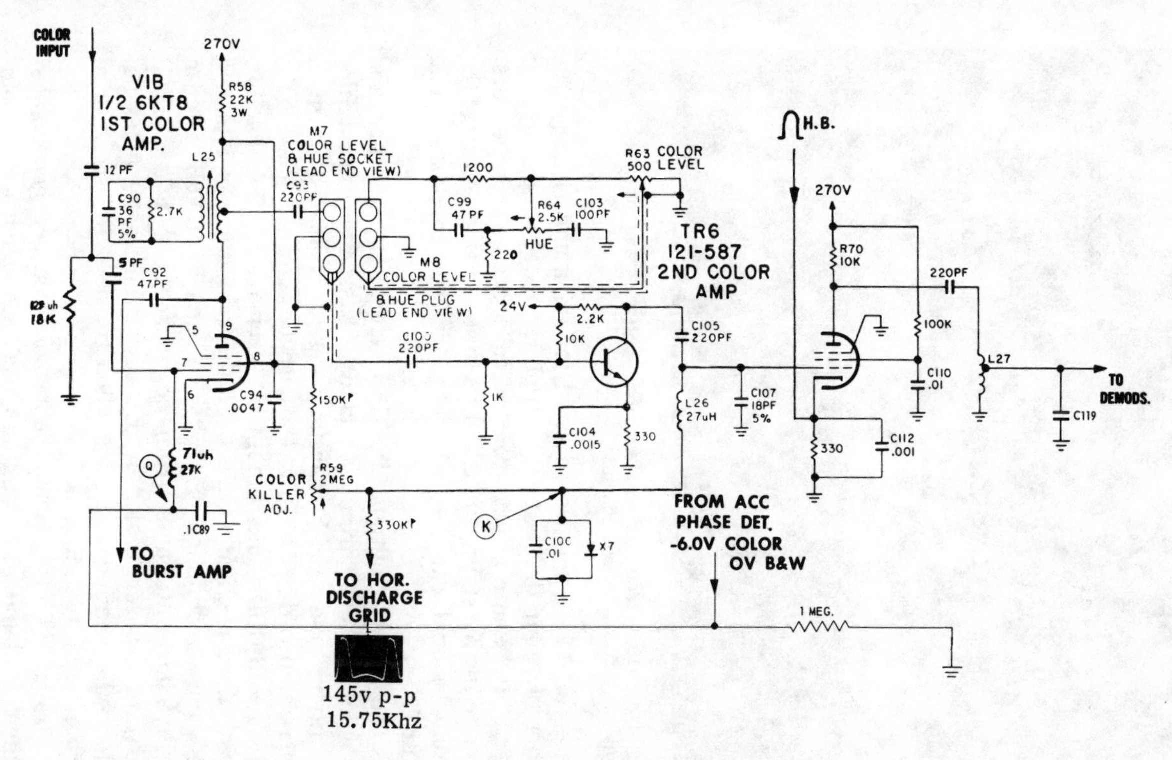

Fig. 9-19. Color amplifier circuit.

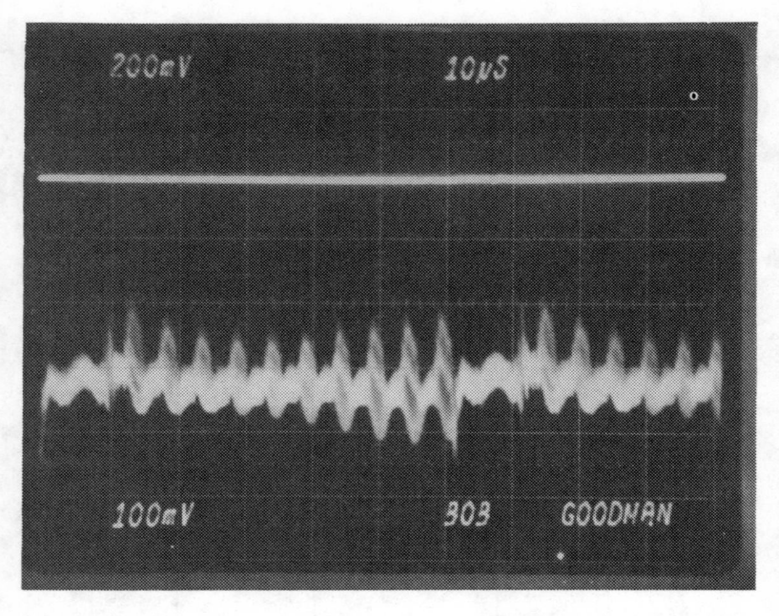

Fig. 9-20. Dual-trace scope waveform—bottom trace normal chroma signal, top trace with signal loss.

plate coil L25, which is adjusted for minimum ringing. The action of L25 in the first color amplifier plate has, by virtue of design, the same effect upon the color channel as the burst input to the burst amplifier. This is to say, that the phase of the chroma signal can be varied by this coil to produce hue control.

The chroma output of V1B is coupled to the hue and color-level control circuitry, and to the base of the second color amplifier, and NPN transistor. TR6 is biased from a zener-regulated 24V supply. The output of TR6 is coupled to the grid of V6B (third color amplifier) through a 220 pF capacitor. Input tuning is nixed by L26, C107, the capacitance of TR6, and the input capacitance of V6B. The cathode of V6B is fed a positive horizontal pulse from the blanker stage to cutoff V6B and provides zero color output during horizontal blanking. Color killer action is also applied to the grid of the third chroma amplifier stage.

Probable Cause: When the loss of color has been isolated to the color bandpass amplifiers, use a scope to trace the point in the circuit where the chroma signal is lost. The dual-trace scope waveform photo in Fig. 9-20 illustrates this technique

The bottom trace shows a normal chroma signal taken at the plate (pin 9) of V1B, while the top trace was scoped at the base of TR6 and indicates a complete loss of the color signal. The fault must then be between these two scope test points. In this case the loss of color was caused by an open C100 capacitor.

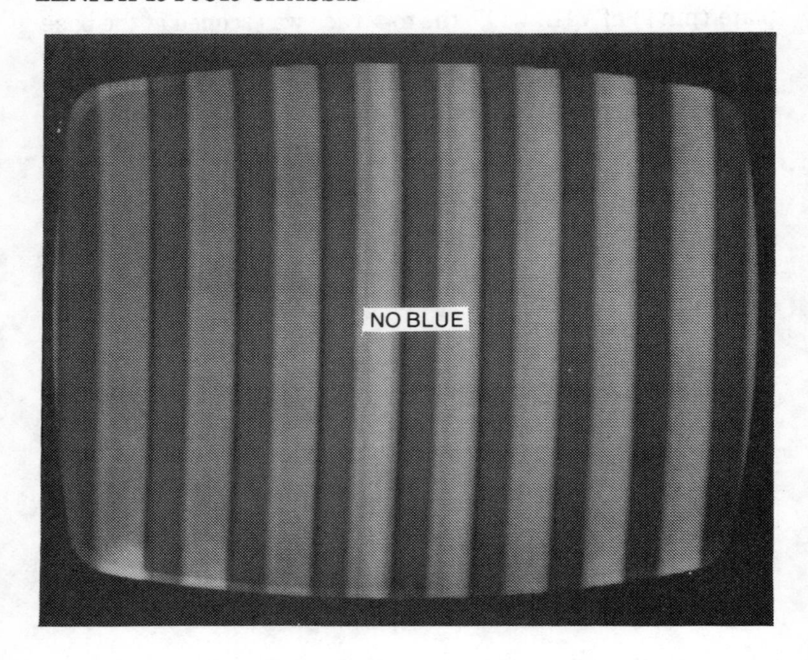

Picture Symptom: Complete loss of blue, but black-and-white picture and sound is good. Look for trouble in color demodulators circuits.

Circuit Operation: The two color demodulator stages (R-Y and B-Y) shown in the Fig. 9-21 are identical in their operation. The 3.58 MHz CW signals appearing at the phase-shift network are coupled to the screen grids of the two demodulators, which operate about 90 degrees out of phase. These CW signals mix with the incoming chroma signal in each demodulator stage and produce the B-Y and R-Y color video outputs. The amplitude of the 3.58 MHz signal at the screen grids of the R-Y and B-Y demodulators is approximately 15 volts peak-to-peak, and this is an important scope check point.

This is a low-level demodulation system and therefore requires further amplification of the detected R-Y, B-Y, and G-Y color signals. The B-Y demodulator output is coupled to the grid of the B-Y color difference amplifier, and the output from the R-Y demodulator is coupled to the grid of the R-Y color difference. A small portion of the B-Y output is coupled to the grid of the G-Y color difference amplifier.

The three 3.9K resistors (R72, R73, and R74) provide equal impedance inputs to the color difference amplifiers. The 1K screen resistors and 220 μF capacitors provide for an amount of 3.58 MHz suppression and are used to bypass the screen grids. The 150Ω resistors in the cathode circuit of each demodulator are unbypassed for stability. You should find approximately 12V P-P amplitude of color information coupled into the control grids of the demodulators. This is a good scope check point for weak or complete loss of color.

Probable Cause: This loss of blue was caused by an open 1K screen-grid resistor for V8A, producing loss of screen voltage at pin 9. For loss of reds check voltages around red demodulator V8B. Use a scope to check for correct chroma signal at pins 7 and 4, then check for the 3.58 MHz CW signal at pins 11 and 2, which should be 90 degrees out of phase with each other.

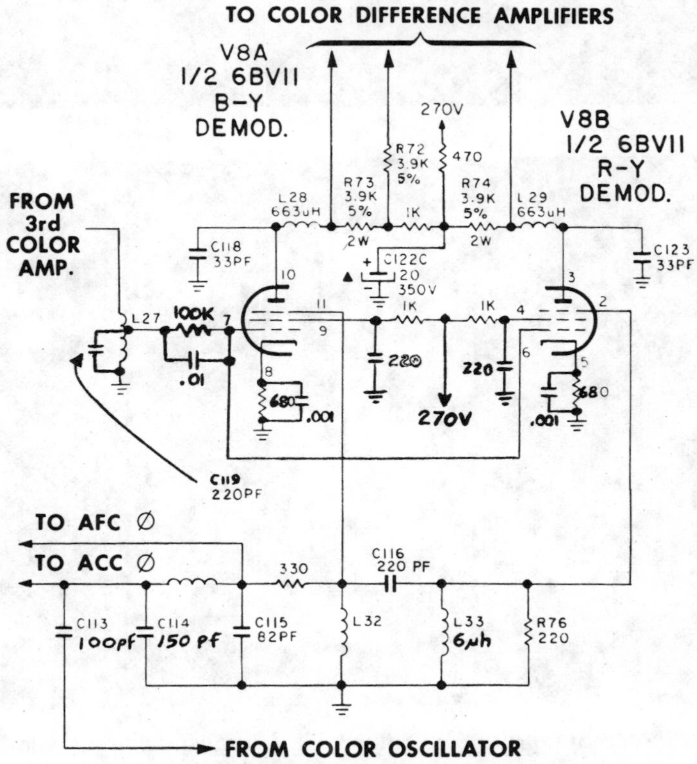

Fig. 9-21. Zenith 15Y6C15, color demodulator circuit.

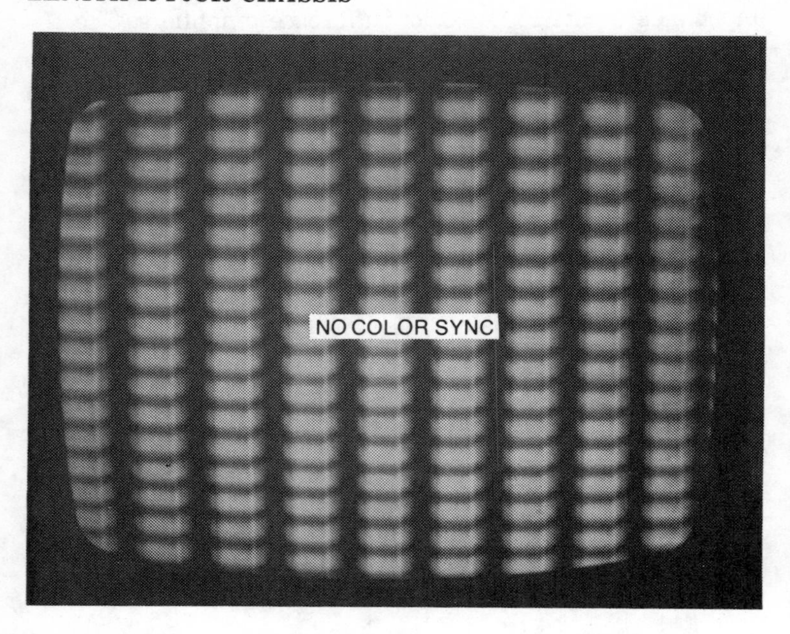

Picture Symptom: No color lock. May also have no color or intermittent loss of color as well as color drop out or delay in color lock-in when changing channels. Look for trouble in the 3.58 MHz oscillator or color AFC phase detector.

Circuit Operation: Refer to the AFC phase detector circuit in Fig. 9-22. Test point W and the 3.58 MHz CW oscillator are involved in a feedback type of operation. The 3.58 MHz signal from point 6 (plate) of the 6GH8A oscillator goes into the phase delay line, where a signal is sampled at the junction of C115 and the 330Ω resistor, to the junction of the two AFC phase detector diodes, X12 and X13. The diodes compare the phase of the 3.58 MHz CW with the burst, developing a DC potential at the junction of the 2.2-meg output resistors (test point W), which then goes to the 6GH8A (V15A) reactance control triode to keep the 3.58 MHz oscillator locked in. This is the complete AFC feedback system.

A properly operating AFC circuit will measure about zero volts with a VTVM at test point W under a no-signal condition. Some possible voltages in excess of one volt at test point W under such conditions can be caused by the following faults:

- A defective AFC phase detector (one diode may be conducting much more than the other).
- A faulty 3.58 MHz CW oscillator and control tube (makes oscillator run off frequency).
- An open L45 burst plate transformer.
- An open or leaky section in one of the dual 0.001 μF capacitors in the AFC phase detector circuit.
- A faulty 42 μH choke (L43) at test point W (this choke is needed for oscillator stability).
- Component leakage or value change in the anti-hunt network (from test point W to ground).
- Incorrect setting of 3.58 MHz coil L47 (check 3.58 MHz oscillator for zero beat).
- The 2.2 meg resistors may have changed value in the AFC phase detector circuit (these are a matched pair).
- Check the 2.2K series resistor in the control grid circuit of V15A for an open or increase in value.

Probable Cause: When test point W was grounded the 3.58 MHz color oscillator would zero-beat, but when ungrounded the oscillator would go way off frequency. This indicates trouble in the AFC control system, which in this circuit was caused by a shorted X12 AFC diode.

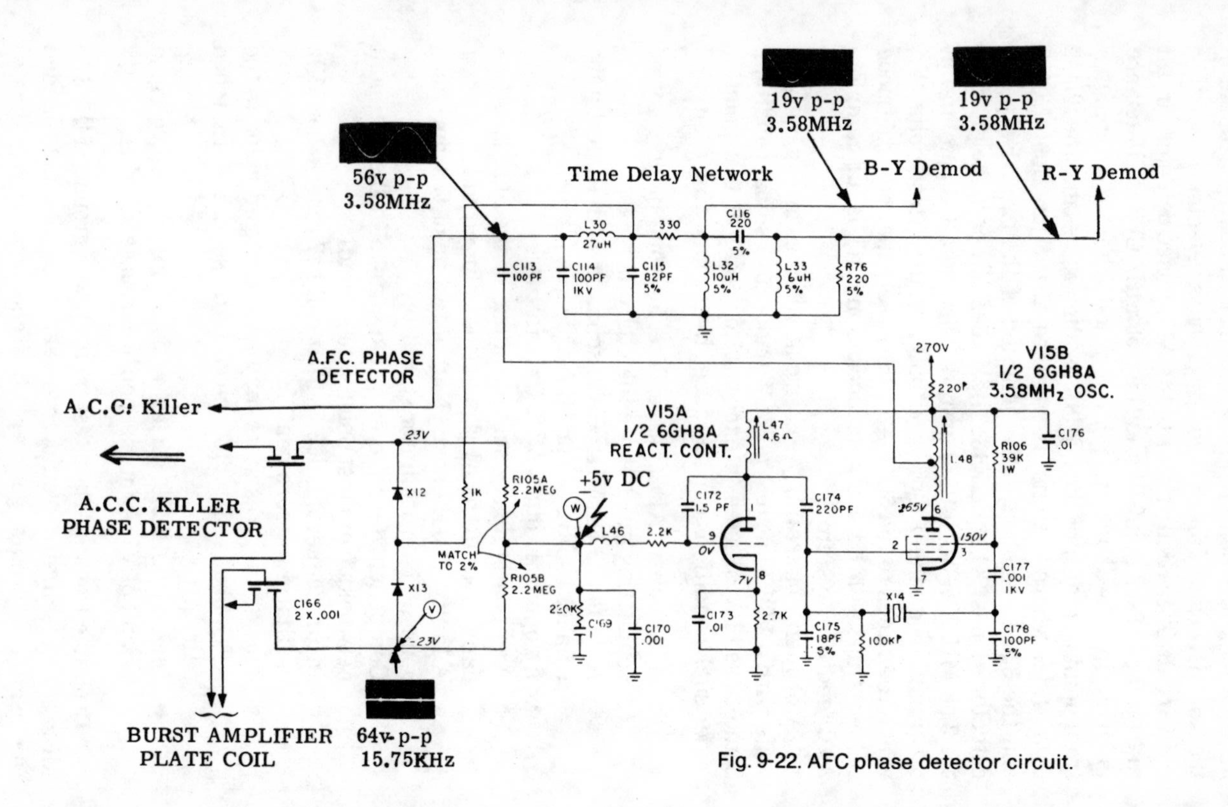

Fig. 9-22. AFC phase detector circuit.

ZENITH 20Y1C38 CHASSIS

Picture Symptom: The picture will develop what looks like jail bars. Another picture symptom that may appear in this same circuit system is a dark area across the top of the picture as shown in Fig. 9-23. Look for such faults in the vertical and horizontal blanker circuits.

Circuit Operation: Vertical and horizontal blanking pulses are applied to the video and color amplifiers (in some sets to the control grids or cathodes of the CRT) in order to remove retrace scan lines. The dual-trace scope is a natural for checking the phasing (timing) of the horizontal blanking pulse and the horizontal sync pulse. If a defective component has changed the phase, the blanking could occur at the wrong place on the picture or be ineffective.

Looking at the simplified blanker circuit in Fig. 9-24, we see that a negative vertical pulse is fed to the video driver transistor. This 9V P-P vertical pulse is coupled through a 470Ω resistor and diode to the emitter via a 2.2K resistor. Diode X1 provides coupling of the pulse to the emitter. The voltage divider consisting of the 150K and 470Ω resistors provide a small positive bias voltage to prevent conduction of

Fig. 9-23. Top of screen dark.

the diode during the sawtooth portion of the pulse and prevents any raster shading.

For horizontal blanking, a negative pulse is coupled through a parallel RC network and then through diode X3 and a 2.2K resistor to the emitter. Note that there is some ringing between the horizontal pulses (5 or 6 sine waves) and should

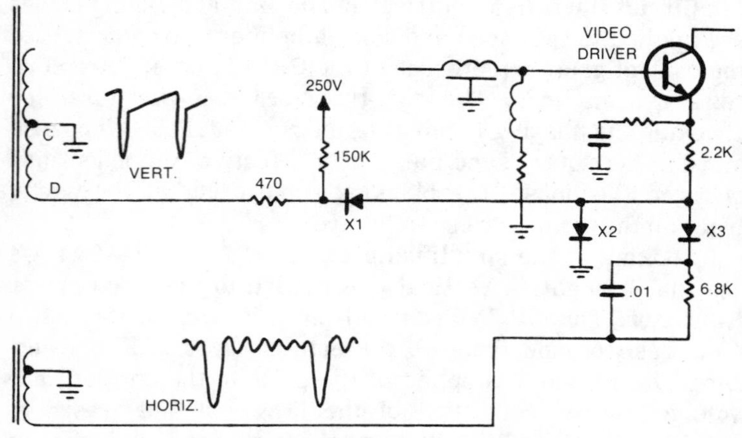

Fig. 9-24. Blanker circuit.

these reach the video driver emitter due to a shorted X3 diode, the picture will produce "jail bars" or drive lines. The ringing is caused by the tertiary windings of the sweep transformer.

To check the timing of the blanking pulse, place one probe of the dual-trace scope at the emitter of video driver and the other probe at the delay line. The blanking pulse should be in time (line up with) the horizontal sync pulse as shown in the Fig. 9-25.

Probable Fault: The jail bar effect is caused by a shorted X3 diode. Also, the 150K resistor connected to the +250V line may decrease in value and cause diodes X1, X2, and X3 to breakdown. When these diodes short out, this will then burn up the 150K resistor and may also overheat the 470Ω resistor. If the 470Ω resistor goes down in value (say to 10Ω), as in Fig. 9-23 this will cause a dark area to appear at the top of the screen.

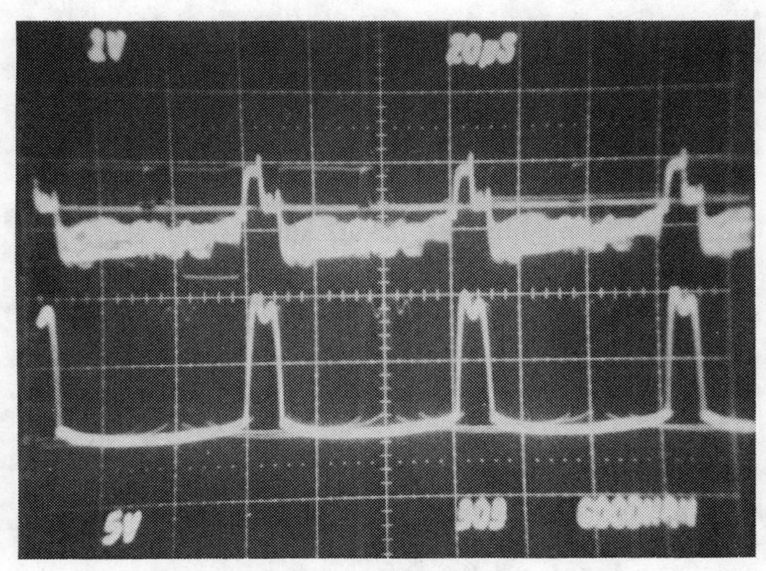

Fig. 9-25. Proper phase of sync and blanking pulses.

Picture Symptom: Color picture was blue with all reds and greens missing. But with the color-level control turned down, the black-and-white picture was good. This color fault could be caused by a defective IC demodulator or incorrect signals being fed into the chip.

Circuit Operation: Let's look at the chroma amplifier circuit shown in Fig. 9-26 to see what caused the wrong color reproduction. The chroma (color) information is coupled to the grid of first color amplifier 6MV8 (V201B) via a 8 pF capacitor (not shown) from the 4.5 MHz trap (test point C1); shunt peaking is provided for equal sideband amplification and thus enhances a better phase response. Test point Q (grid of the first color amplifier circuit) is returned through a 220K resistor to ground, to complete the ground path return.

The color signal is amplified by V201B, but is reduced some by first color amplifier plate coil L213, which is tapped to provide impedance matching to IC 221-62. The chroma output from V201B is coupled through C262, L214, and a diode-resistor network circuit. The chroma signal information is then

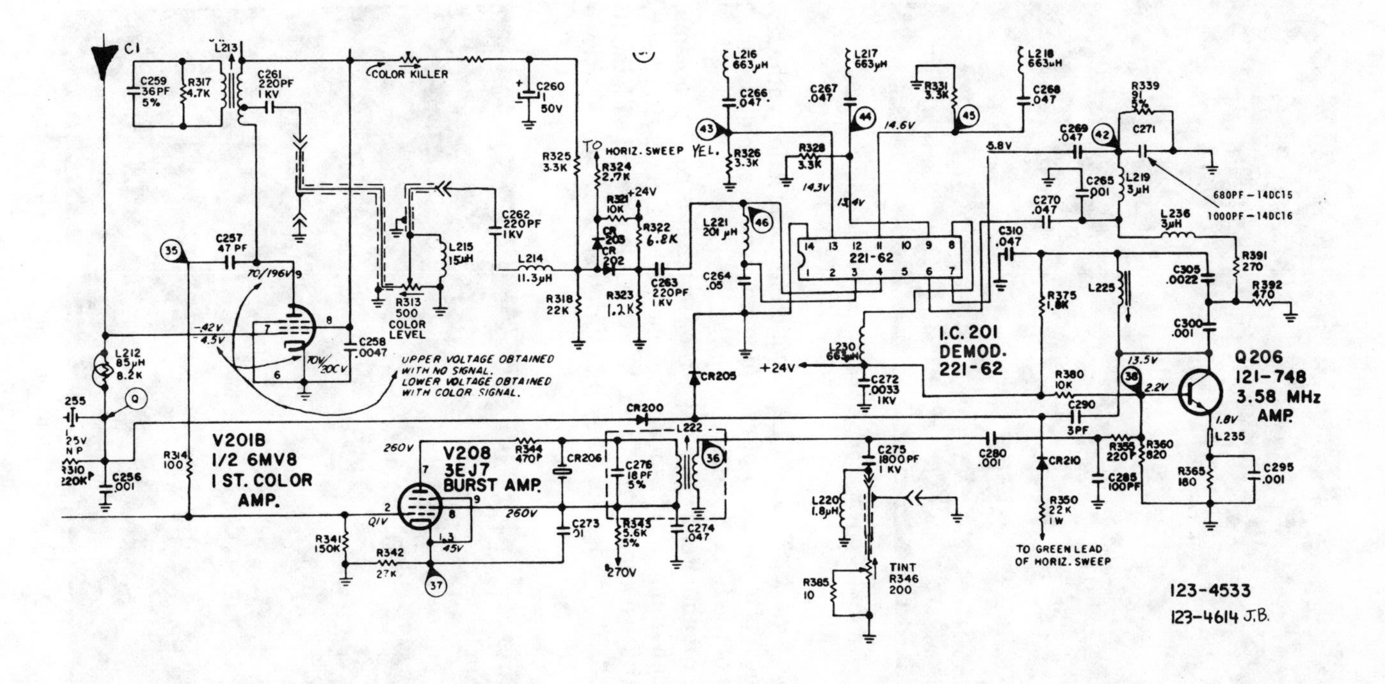

Fig. 9-26. Chroma amplifier and demodulator circuits.

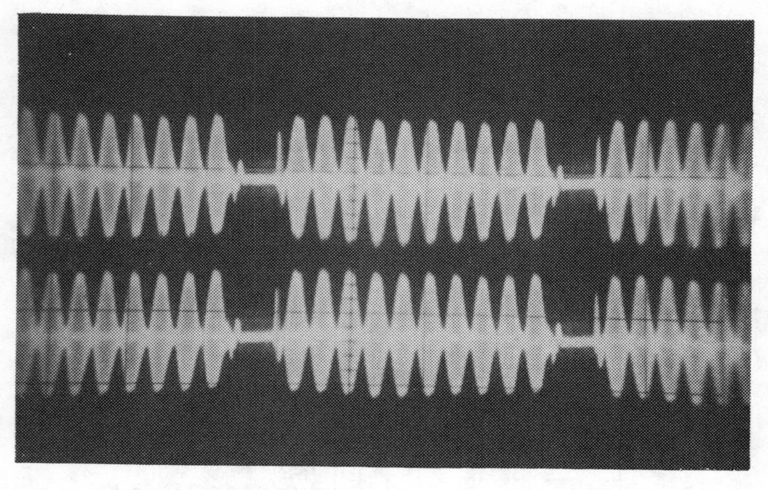

Fig. 9-27. Correct scope signals at pins 3 and 4 of IC 221-62.

coupled via C263 to coil L221, which in conjunction with C264 will now provide an in-phase balanced chroma signal for pins 3 and 4 of the IC demodulator chip.

Probable Fault: For any wrong or missing color problems, the demodulator IC should be checked out first. To check, substitute a known good chip. If no trouble is found with the IC and all DC voltages are near normal, then use a scope to trace

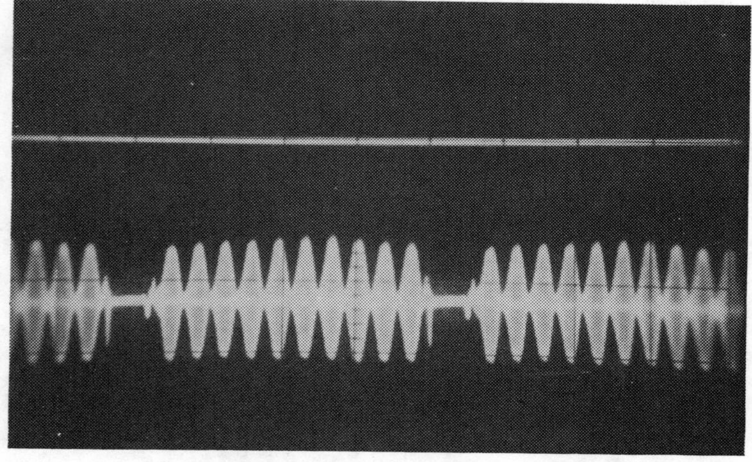

Fig. 9-28. Chroma signal missing at pin 3 of the IC

the chroma signal from control grid of V201B to point 3 and 4 of the IC. With a dual-trace scope the two chroma signals at pins 3 and 4 of the IC should appear as shown in Fig. 9-27. In this chassis, one of the chroma signals was missing at pin 3 of the IC as noted in the Fig. 9-28. This signal loss was due to a shorted C264, a 0.05 μF capacitor.

Picture Symptom: Overall picture has a blue cast and may have red and blue streaks across the screen. There is a definite loss of color, and this may be of an intermittent nature. Look for trouble in the color-burst ringing and 3.58 MHz amplifier circuits.

Circuit Operation: The burst amplifier stage (Fig. 9-29) separates and amplifies the burst signal and provides a 3.58 MHz CW signal for crystal ringing. The burst signal at the plate of V208 rings a high-Q circuit consisting of crystal CR206 and a tuned circuit. The incoming burst signal is used to develop the 3.58 MHz CW signal, so an oscillator stage is not required. Also, since the 3.58 MHz signal is automatically locked in by virtue of design, an AFC circuit is not required.

The ringing (oscillatory effect) of the crystal and tuned circuit is sustained over a complete horizontal sweep cycle. Because of a high circuit Q, the damping (or fall-off) of the ringing signal is very small and the crystal continues to produce a 3.58 MHz signal till the next burst signal.

Since the ringing circuit is a high-Q type, a scope probe should not be connected at the burst amplifier plate circuit to

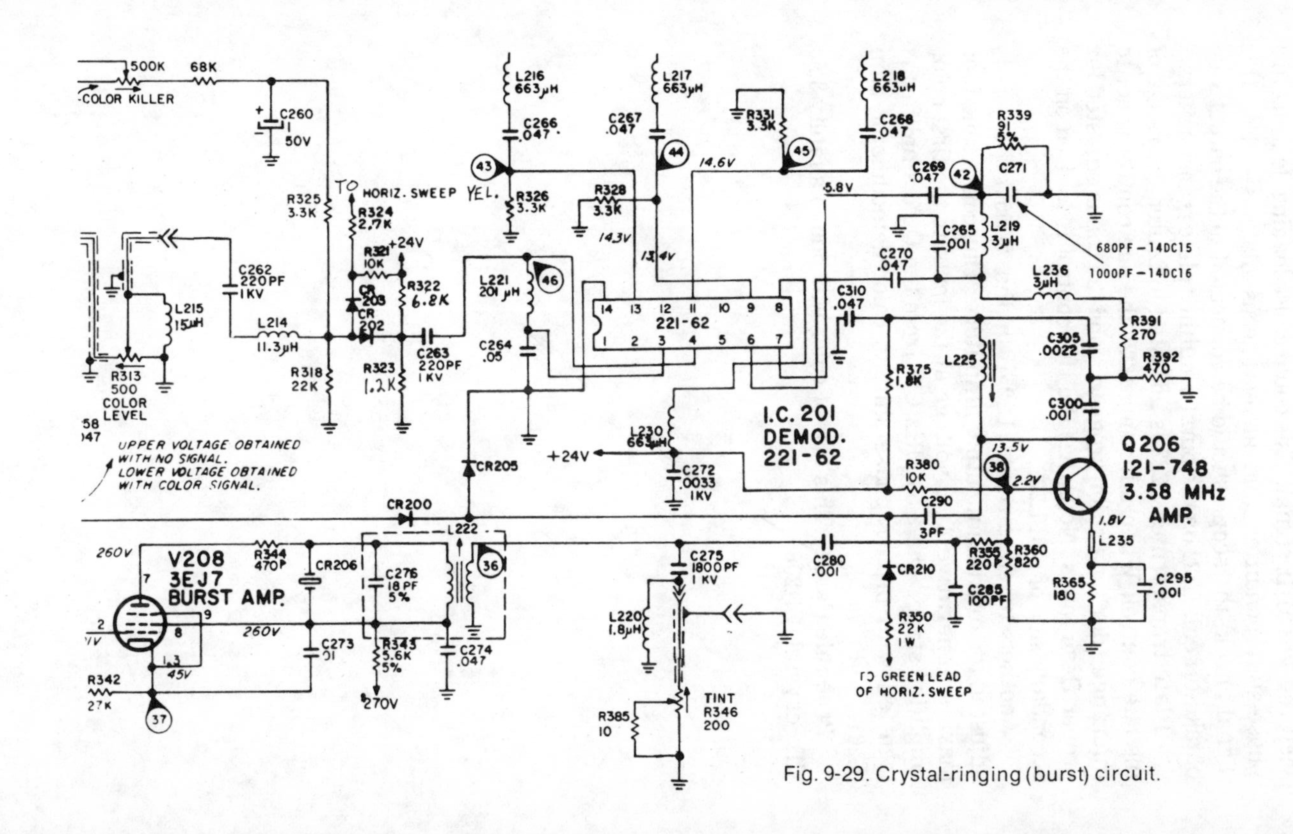

Fig. 9-29. Crystal-ringing (burst) circuit.

view the 3.58 MHz signal. This could cause loading, leading to a phase-shift condition and an erroneous diagnosis. For this signal check, the scope should be connected to the output side of take-off coil L222 on the burst amplifier plate transformer.

Plate transformer L222 is adjusted to center the range of the hue control. When adjusting L222 the hue control is set to midrange (during a color program) and the coil is adjusted for proper flesh tones. Note that the tint control is located on the secondary side of coil L222.

Another tuned circuit (L225) in the Q206 collector is adjusted for maximum output amplitude. This coil is tuned for maximum negative ACC voltage at the input to the first color amplifier stage. The 3.58 MHz CW from the Q206 amplifier is then fed with proper phase shift to color demodulator chip IC201.

Probable Cause: This loss of color was due to a faulty 3.58 MHz crystal, CR206.

Brightness and Contrast Problems

The problems listed in this section are not necessarily troubles with the brightness or contrast control, though some faults may be found in these control circuits. Here we describe symptoms where the picture itself seems excessively dark, is too bright, lacks contrast, etc.

The photos in this chapter may not always show the specific symptom that you are experiencing, since varying degrees of the symptom may occur in any given set. The picture symptom descriptions below the pictures try to express the extremes of the symptoms and to note where alternate or intermittent problems may manifest themselves.

In most cases the circuit operation is described in sufficient detail to permit you to track down and pinpoint the defect. And though different manufacturers may take slightly different approaches in circuit design, the problem symptoms usually originate from the same general circuits regardless who makes the set. The most significant difference is likely to occur between sets having direct-coupled and capacitor-coupled amplifiers, since DC amplifiers can easily cause changes in brightness through changes in bias level occuring through a fault in any stage of the amplifier. AC amplifiers, on the other hand, are more likely to produce contrast problems.

Picture Symptom: The customer complaint for this set was a blank screen and no sound with buzz. A good place to look for these type symptoms would be in the power supply. This chassis employs three scan-derived power supplies producing 33V, 24V, and 235V DC.

Circuit Operation: Positive horizontal pulses from terminal 10 of the sweep transformer (Fig. 10-1) are coupled through fuse F1000 to the anodes of diodes D1000 and D1001. The pulse is rectified by D1000, filtered by C102B, C102C, and R109 to become the B+ 33V supply. D105 is a protection diode that carries any negative transients to ground, thus protecting the solid-state components connected to the supply.

The same pulse is rectified by D1001 and filtered by C1000, producing a 35V supply which is dropped by voltage regulator IC1000 to become the B+ 24V supply. D1002 and D1003 are for transient protection, working in the same manner as D105. Further filtering of the B+ 24V line is provided by C1001 and C1002. IC1000, the B+ 24V regulator, maintains this supply at 24V during normal load variations.

The third scan-derived power supply is provided by rectifier D100. A positive horizontal pulse from terminal 6 of the sweep transformer is coupled to D1000 by R101. This

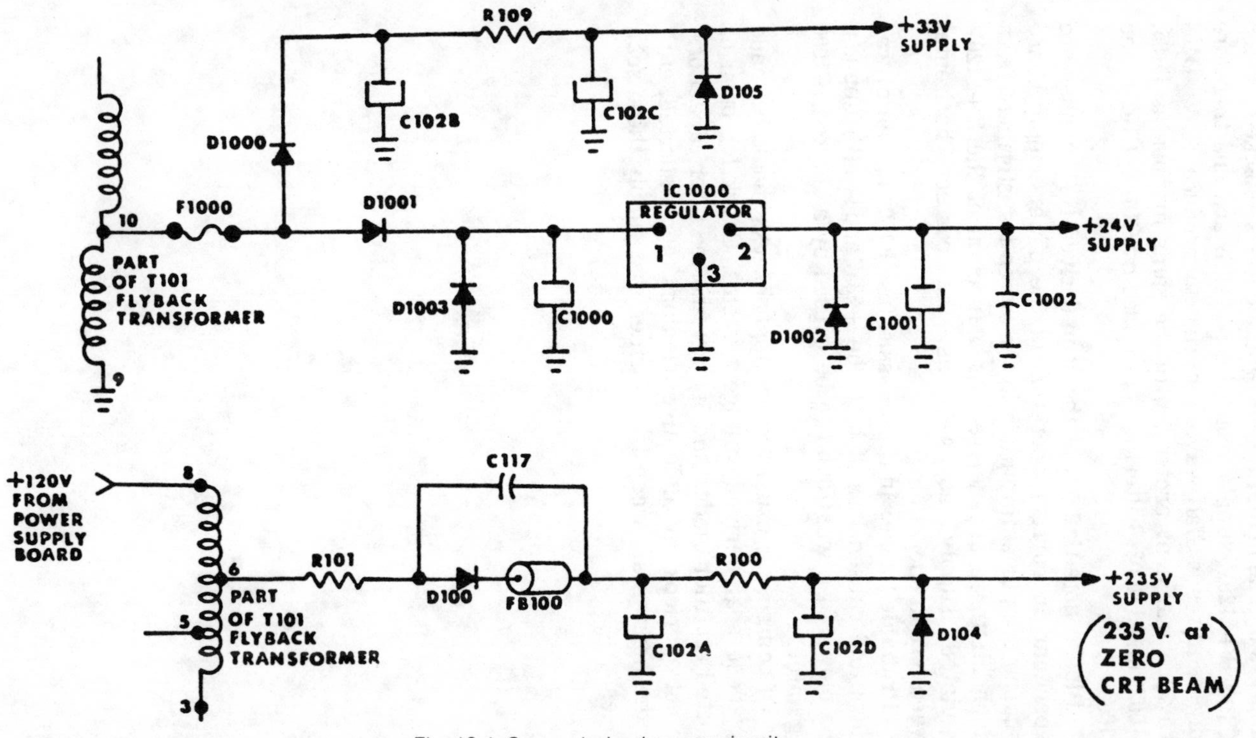

Fig. 10-1. Scan- derived power circuit.

supply is unique in that the horizontal pulse is "stacked" on top of the B+ 120V supply from the power board and therefore only requires a half-wave rectifier to produce 235V. D104 provides transient protection in the same manner as D105, D1002, and D1003. Filtering is provided by C102A, C102D, and R100.

Please note that all of the scan-derived B+ supplies are dependent on correct operation of the power supply board because the pulse amplitude in the sweep transformer (from which all three are derived) is a function of the B+ 120V regulated supply as well as the correct horizontal sweep-circuit operation.

Probable Cause: In this chassis, fuse F1000 (1.5 amp) was blown and caused loss of both the +33V and +24V supplies. The blown fuse was found to be caused by a shorted IC1000 regulator.

Of course, the usual power supply checks should be made for these scan-derived voltage systems. Check for open or shorted rectifier diodes and open or shorted filter capacitors. Do not overlook any defective windings (open, shorted, etc.) on the horizontal sweep transformer that supply these scan pulses.

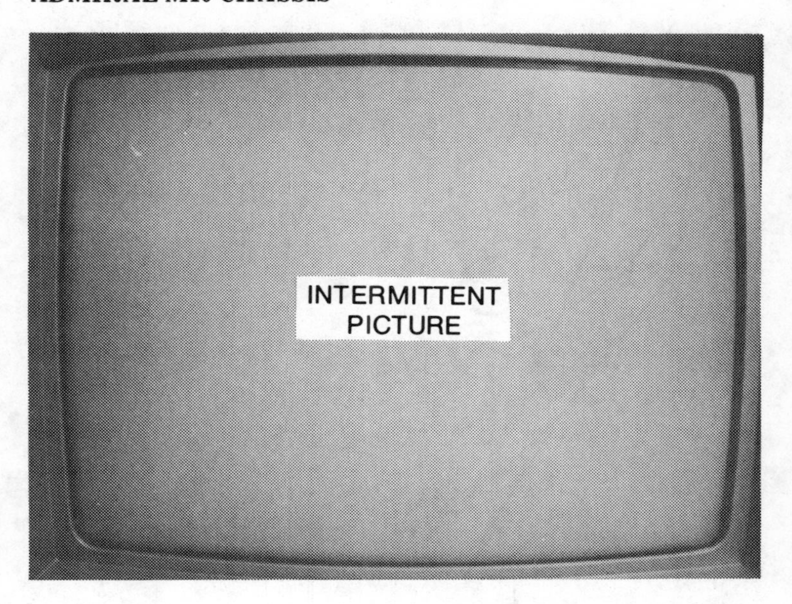

Picture Symptom: Pulsating audio and raster flashing on and off at one second intervals. Look for trouble that will cause the high-voltage shutdown circuit to be activated. The HV shutdown circuit is a combination protection and warning device. It samples the horizontal pulse amplitude present at the sweep transformer, sensing too much horizontal drive or a HV condition, and produces the pulsating audio and flashing raster.

Circuit Operation: The high-voltage shutdown circuit in Fig. 10-2 shows that a negative-going horizontal sample pulse from the sweep transformer is coupled to the cathode of zener D804 via C114 and R111. D108 provides negative-tip clamping and maintains the negative peak at 0.6V below ground potential.

Should a fault occur that allows the HV to increase, the amplitude of the sample pulse increases in the positive direction, driving the 27V zener into breakdown and applying the signal to the gate element of SCR Q805, causing it to conduct. This turns off the horizontal oscillator by grounding Q801's base through D805 and the SCR. With the oscillator turned off, no current flows in the horizontal driver or sweep output stage.

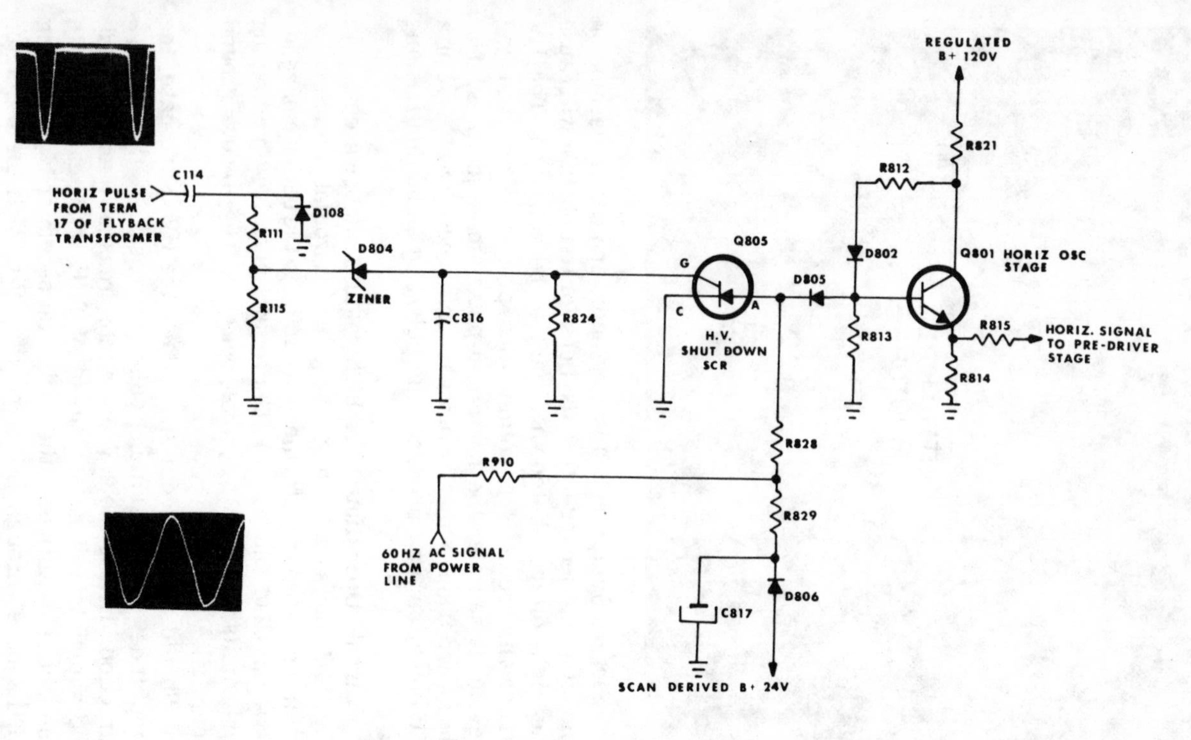

Fig. 10-2. Admiral M10 chassis, HV shutdown circuit.

Capacitor C817 (50 μF) is initially charged to 24V minus the drop across one diode. When Q805 conducts, its anode voltage drops and the base of Q801 is grounded. The 24V supply then drops to zero since this is a scan-derived power supply. C817 discharges through R829 and R828 and the conducting SCR. An AC voltage, derived from the high side of the AC line, is injected at the junction of R829 and R828 via R910. The peak value of the AC at this point is 2.6V. When C817 discharges to 3.6V, the anode voltage drops to zero on the negative peaks of the AC. This occurs approximately one second after the SCR conducts. When the anode voltage drops to zero, the SCR turns off and the horizontal oscillator restarts. If a circuit defect still exists, the SCR is turned on and the cycle repeats itself. The pulsating scan-derived B+ voltages to the audio circuits result in a "putt-putt" sound from the audio circuit at the rate of once per second.

Shutdown circuit checks:

1—Check the B+ voltage. If necessary, adjust accurately to +120V with control R901. Be sure to make this adjustment with the CRT at zero beam current.
2—Set the tuner to an unused channel.
3—Set the brightness and contrast to minimum (zero beam).
4—Slowly turn the horizontal oscillator adjustment (T800 core) counterclockwise until the shutdown circuit triggers. (Bend the tabs on the top shield of the horizontal oscillator coil to permit full rotation of the adjustment rod.)
5—Disable the shutdown circuit by clipping a short jumper lead across R824, a 6.2K resistor.
6—Measure the high voltage. It should be 29 to 33 kV.
7—Remove the temporary jumper wire from R824.
8—Readjust T800 for correct horizontal frequency, and check for a good lock on all active channels.

ADMIRAL M10 CHASSIS

Picture Symptom: No picture or sound. Loss of +120V regulated supply. Check for short in the +120V line or trouble in the power supply circuit.

Circuit Operation: This power supply uses a half-wave rectifier circuit (Fig. 10-3) with the low side of the power line at chassis ground. A series B+ regulator circuit compensates for line voltage and load current variations with three transistors. Two are series-pass stages and one a pass-driver stage. The entire regulator circuit is voltage referenced to zener D901.

D900 is the rectifier diode. Filtering of the unregulated B+ supply is provided by C101A. Resistor R116 serves as a bleeder resistor for the supply when the set is turned off. R907, an NTC thermistor, limits the in-rush current to a safe level when the set is first turned on. The unregulated B+ is fed to the collectors of the parallel-connected Q101 and Q102 pass transistors.

The scan-derived 235V B+ is applied to the cathode of reverse-biased D901 via R900 and R909. R901 and R902 constitute a DC voltage-divider network to set the operating point of Q900 and thus the entire regulator circuit.

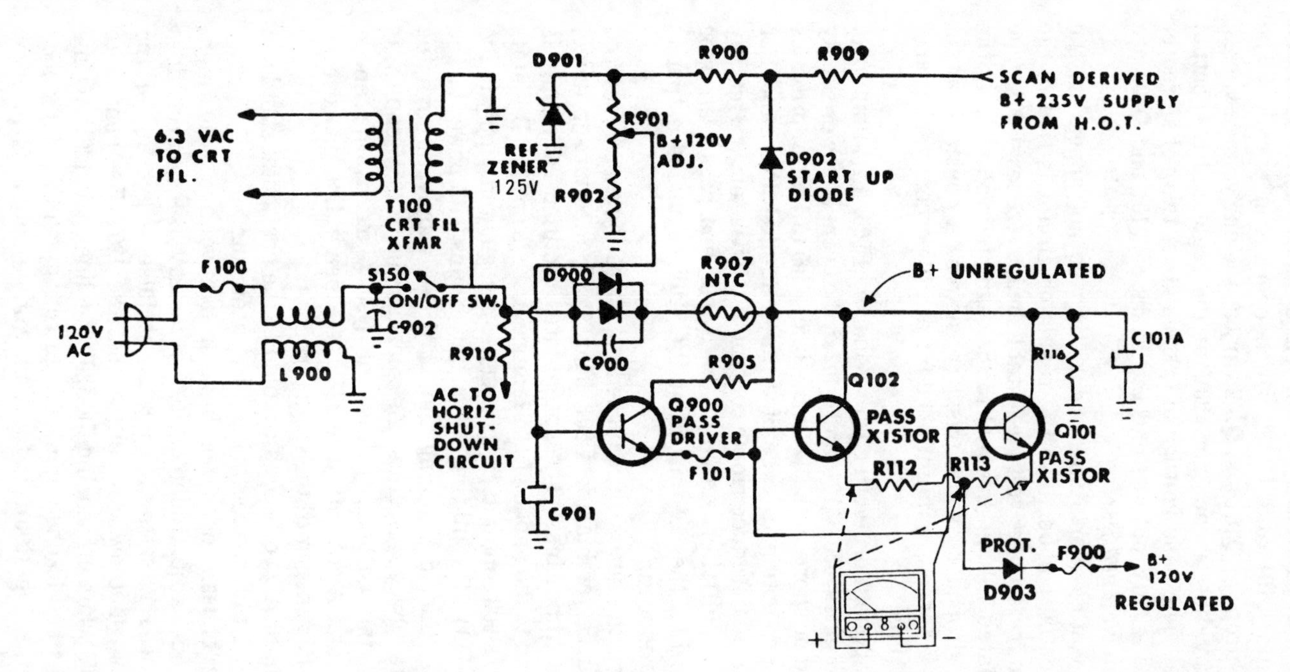

Fig. 10-3. Admiral M10 chassis, power supply circuit.

The unregulated B+ supply is dropped across Q102 and R112, and Q101 and R113, which operate in parallel. The voltage drops at Q101 and Q102 vary as the line voltage varies, thus providing a constant 120V at the regulated B+ output terminal. The base voltage of Q900 is about two volts greater than the +120V output, and the bases of Q101 and Q102 are about one volt greater.

As voltage variations occur, they do not appear at the regulator output because the regulator circuit is tied to the constant reference voltage established by zener D901. The unregulated voltage is equal to the regulated voltage plus the voltage drop across the pass transistors. A constant output voltage is therefore maintained for a ±10 percent variation of the 120V line.

Component protection is provided by F900, a 1.5-amp fuse. F101 protects Q900, Q101, and Q102 in the event of power supply overload or short on the 120V B+ line. F100 (3.1 amp) provides overall protection in case of a major fault. Filter capacitor C901 is used to reduce ripple content at the base of Q900 and, by capacity multiplier action of the driver and pass transistor combination, the B+ ripple at the 120V regulated output is greatly reduced.

When the set is first turned on, there is no scan-derived 235V source for biasing D901. Thus D902, the start-up diode, provides initial bias. The unregulated rectified DC supply is connected to the anode of D902. Since there is no 235V B+ at its cathode, D902 conducts, applying the unregulated supply to the cathode D901. Regulating action now begins with power on the 120V B+ line, developing the 235V B+ supply through the horizontal sweep circuits. As soon as the +235V supply is present, the cathode voltage of D902 exceeds the anode voltage, and it cuts off. The 235V B+ then maintains reverse bias on the zener until the set is turned off.

A small AC voltage is coupled to the horizontal shutdown circuit by R910 to meet its input requirements.

This chassis does not have a high-voltage adjustment control but, adjustment of R901, the 120V B+ control, does affect the HV setting. To prevent false triggering of the horizontal shutdown circuit, adjust the 120V B+ control to read 120V when measuring the 120V B+ line (with 120V AC line input) instead of checking the high voltage on the CRT anode while adjusting R901. If the high-voltage supply is in error when the B+ line is 120V, there is probably a fault in the high-voltage circuitry.

The following test will indicate whether or not the pass transistors are functioning properly. Use a VOM with a 1.5 VDC scale, or a VTVM without its negative terminal referred to ground. **CAUTION**: Disconnect power from the set when connecting the negative meter lead, for accidental grounding of the regulator circuit will blow fuse F101 and may damage Q101 and Q102.

1—With power off, connect the negative lead of your VOM to the junction of R112 and R113 (on terminal strip, in upper-left corner of chassis).
2—Turn set on the adjust for normal picture brightness.
3—Touch the positive probe of your meter to the other end of R112. Your meter should read 0.5V to 0.6V.
4—Touch the positive probe to the other end of R113. Note the meter readings.
5—The difference in voltage across R112 and R113 should not be greater than 0.1V. If the voltage difference exceeds this amount, one of the pass transistors is not operating properly.

Probable Cause: This receiver's trouble was caused by a shorted pass transistor (Q102) that also blew out fuse F900 and caused a loss of 120V B+.

Picture Symptom: No vertical sweep, or a dark screen. May also blow fuse F910. Look for trouble in the scan-rectifier circuits located on the power-supply horizontal-buffer module.

Circuit Operation: The scan-rectifier circuits (Fig. 10-4) supply voltage to the vertical module. To develop 22 volts, Y914, C916, Q900, and Q902 are used. A double-ended supply consisting of Y642, Y646, C648, and C649 supplies −13V and +13V for the vertical module. The windings for this supply are connected to terminals 8, 9, and 10 of the horizontal sweep transformer. Another supply consists of Y920 and C922, and the negative end of this supply is connected to the +140V half-wave supply, which allows a lower voltage filter capacitor to be used.

The rectifier diodes used in scan-rectified circuits conduct about 80 percent of the time, and they must therefore be made for long duty-cycle operation. In addition, they must be able to withstand the fast rise time, high-amplitude, retrace pulse. For this reason, exact replacement parts must always be used.

Probable Cause: The voltage derived from horizontal scan rectifier Y914 is fed to the collector of Q615 via the vertical size control, which is located on the vertical module. In this set, the loss of vertical sweep and blown F910 fuse was caused by a

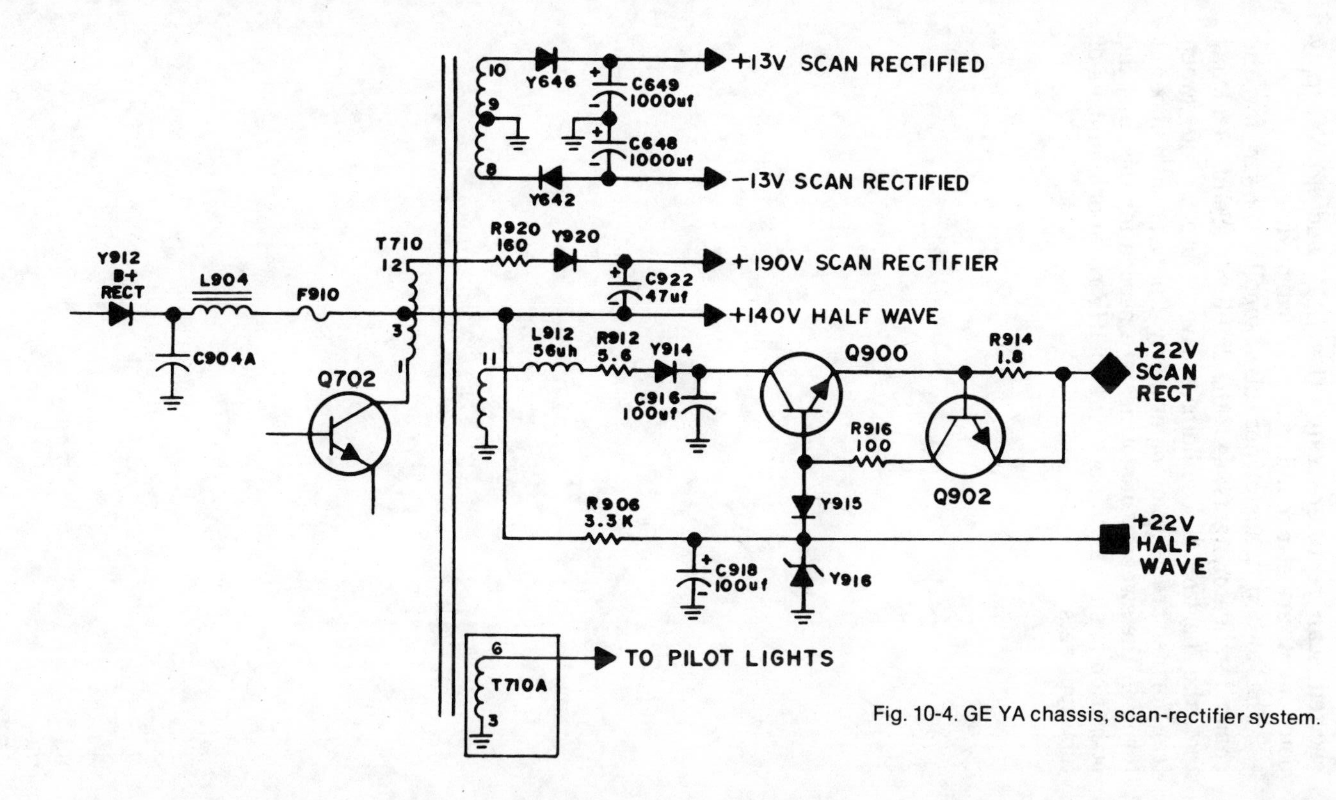

Fig. 10-4. GE YA chassis, scan-rectifier system.

shorted scan rectifier, Y914. The −13V and +13V supply components are also found on the vertical module.

The sweep transformer also supplies power for the channel indicator lights, so if the dial lights are lit, you know that the horizontal oscillator, driver, output, and sweep transformer are functioning properly. Only a small number of turns on the sweep transformer are needed for this, and the operation is similar to powering the HV rectifier filaments of tube-type sets.

Picture Symptom: The picture would slowly fade out, after a warmup period, to a very bright, blank screen with retrace lines. There could also be an intermittent picture fade-out condition. Look for trouble in the video amplifier.

Circuit Operation: The video amplifier (Fig. 10-5) contains two sections, which are coupled by capacitor C148. In the first section the emitter of Q104 provides a low-impedance signal source for the chroma circuits and Q106. The collector of Q104 provides a composite video signal for sync buffer Q105, and this collector signal should be three times the amplitude of the base signal. The gain of Q106 is controlled by contrast control R164, which varies the AC degeneration in the emitter circuit. At maximum contrast setting, the collector signal amplitude of Q106 should again be three times the amplitude of the base signal. Y106 is the clamp diode for DC restoration.

In the second section of the video amplifier, we find second video buffer Q108, third video amplifier Q110, and video driver Q109. The delay line, which is also in series with peaking coils L114 and L116, is connected between the emitter of Q108 and the base of Q110. The emitter of Q109 serves as the output point of the video signal, and the signal is fed from this point to the emitter resistors of the RGB color output transistors. Because

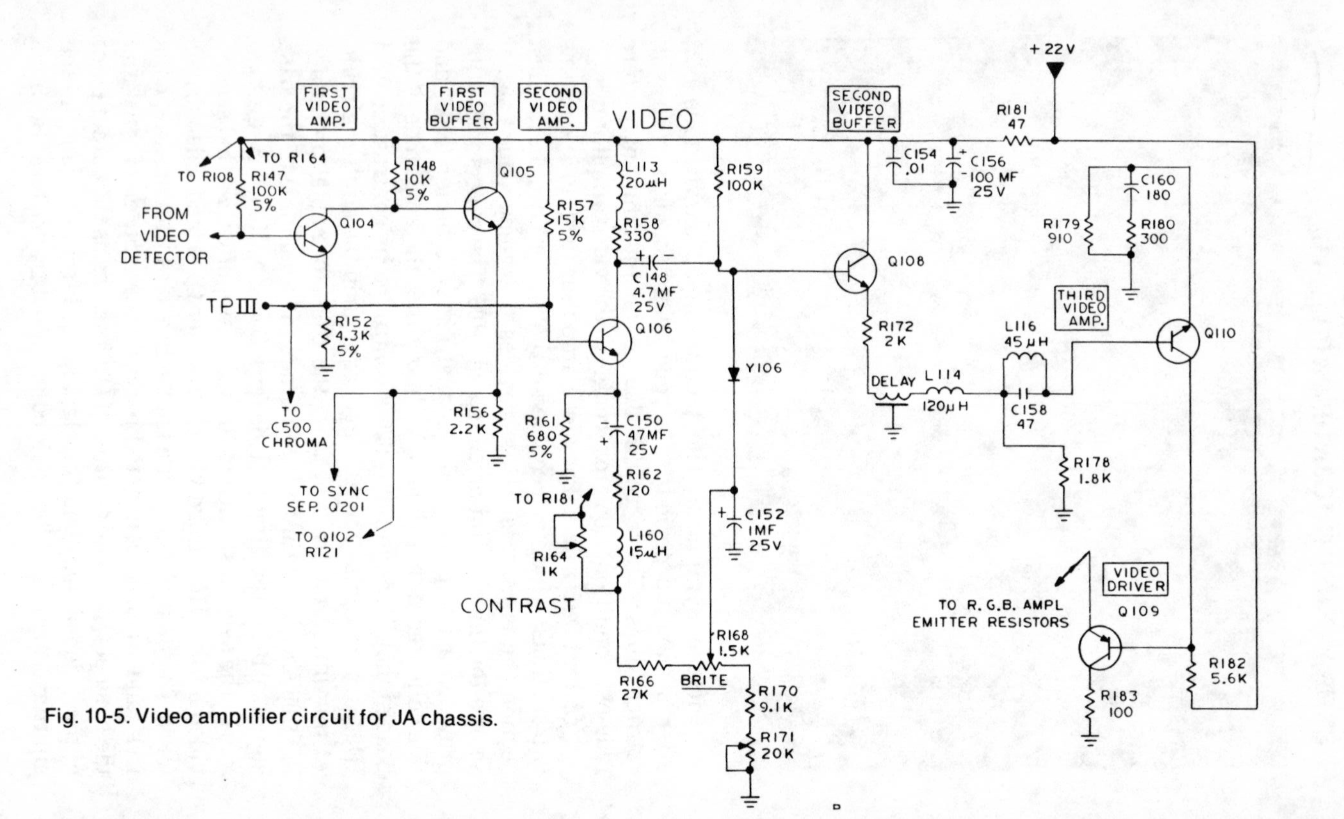

Fig. 10-5. Video amplifier circuit for JA chassis.

this circuitry is directly coupled, a DC voltage change at the base of Q108 will cause a DC voltage change at the CRT, and this voltage change will be amplified and inverted. Thus, if the voltage at the base of Q108 becomes 1.5V more positive, the voltage at the CRT grids will become about 30V less positive. (Color matrixing takes place in the three RGB color output transistors.)

Probable Cause: Use an oscilloscope to trace the video information through the various stages. Because this is a directly coupled amplifier system, voltage readings can be quite misleading. This slow picture fade-out was caused by a defective delay line. The delay line did not open up, but developed a leakage from the ground shield to the inside coil, in effect, shunting the video signal to ground and accounting for the slow fadeout after the set warmed up.

Picture Symptom: No picture or sound. Circuit breaker had tripped. When the circuit breaker was reset and the receiver turned back on, the breaker would trip out again in about 5 seconds. Look for a short or circuit overload in the circuits fed from the B+ voltage supply. Look for trouble in the video output stage (tube V5A).

Circuit Operation: Refer to the circuit shown in Fig. 10-6. The video output stage uses the pentode section of an 8AL9 compactron, designated as V5A in this circuit. The plate of this tube is directly coupled to the three cathodes of the picture tube. Hence, any changes in the direct current flow through the video output tube will cause a brightness change on the screen of the picture tube. Brightness control R194 provides a way for varying this DC flow.

Video gain is adjusted by contrast control R193. Since the high frequencies are bypassed to ground by capacitors 4C176 and C178, the high-frequency gain remains fairly constant at all contrast control settings. This control has maximum effect on video frequencies below 2 MHz.

In order to supply proper plate voltage to V5A and bias voltage to the cathodes of the CRT, +280V is fed via the 220 μH

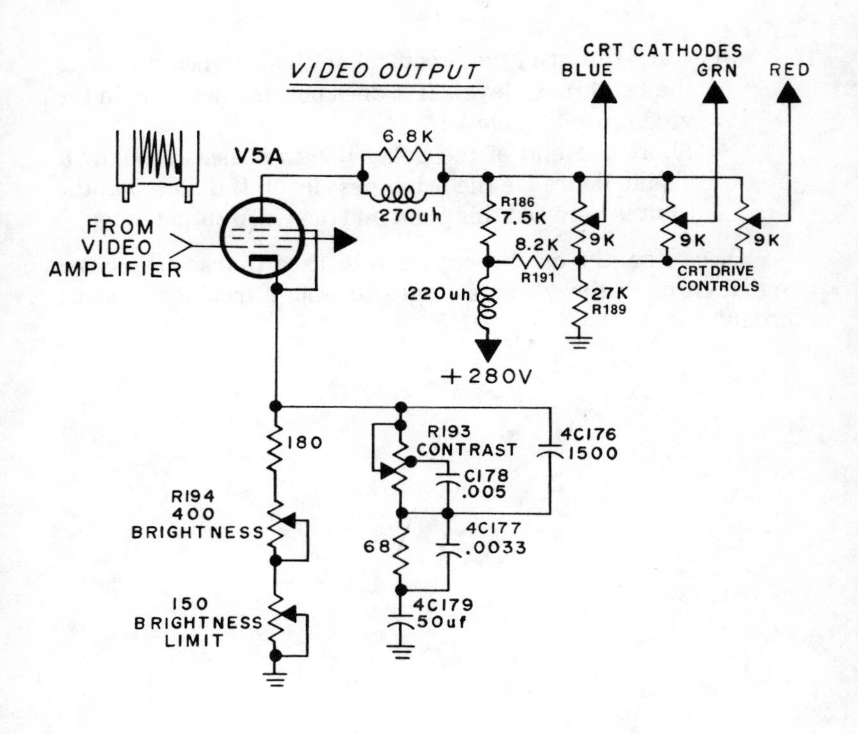

Fig 10-6. GE C-2 chassis, video output circuit.

video choke coil to a voltage-divider network that consists of resistors R186, R191, and R189.

Probable Cause: Voltage-divider resistors R189 and R191 had decreased in value and have a charred look caused by high current flow, possibly due to a short in the picture tube. As the value of these 8.2K and 27K, 2-watt resistors go down, more current is drawn from the +280V B+ supply, and the circuit breaker trips out. The 220 μH coil may also overheat and open too, in which case there will be no plate voltage at V5A and the CRT screen will be black. Should R191 and R189 go up in value (or open up), the picture will appear almost normal, but the CRT drive adjustment controls will have no effect, and the gray-scale tracking for the picture will not be correct.

The directly coupled stages can also be used for quick checks in troubleshooting these video circuits. If there is a raster on the screen but no video, use the following checks to help isolate the defective stage.

- Vary the brightness control setting. It should change the brightness level. If it does not, the defect is in the video output stage V5A.
- Short one end of the delay line to chassis ground. It should decrease the brightness level. If it does not, the fault is between this point and the video output stage.

Of course, the oscilloscope can be used to trace the video signal through the various stages to isolate troubles in short order.

Audio Problems and Remote Control

Because there are no picture symptoms for audio problems, we will only give verbal descriptions. In most cases, the nature of the audio defect provides you with the most valuable clues as to the source of the trouble. Regardless of the make or model of the receiver in question, when audio problems develop, you know where to start looking for the trouble, for it will be somewhere between the sound take-off point and the speaker. A schematic of the chassis you are troubleshooting and a basic knowledge of audio circuits will get you off to a good start.

In other cases, the problem may be related to the audio, but you will not find the answer in the audio circuits. A very important consideration is the B+ supply circuits. Check for correct voltages and proper filtering. Another common example is the complaint of a high-pitched whistle whose frequency is a component of the 15.75 kHz horizontal sweep frequency. This problem may appear in any vintage receiver, though it is indeed more common with tube-type receivers than their transistorized cousins, because of a vacuum tube tendency to become microphonic.

Some remote control circuits are also included in this section along with some troubleshooting tips. Remote control is covered here because virtually all such systems provide for adjusting the volume level in steps, which of course opens the way for many types of audio problems. And while you can override a faulty channel-changer mechanism by manually turning the dial, few remote control systems provide for manual override in the audio sections.

ADMIRAL K20 CHASSIS

Audio Symptom: Loss of sound or audio distortion. No picture symptoms. Look for trouble in the sound system.

Circuit Operation: The integrated sound circuit (Fig. 11-1) used in this chassis performs the functions of 43 components in converting the 4.5 MHz sound-IF to a usable audio. The IC is located on the transistor board (Fig. 11-2) next to discriminator transformer T202. The IC is of the plug-in type and no soldering is necessary.

The 41.35 MHz sound-IF and 45.75 MHz video-IF detector sources are derived directly from the collector of the third IF transistor through C201. C201 and L201 resonate at about 42 MHz. Thus, the two carriers (sound and video) appear about equally apart on the sides of the response curve at this point.

The detected 4.5 MHz beat between the sound and video carriers is passed from sound detector D201 to the primary winding of T201 (4.5 MHz sound-IF input transformer) via divider network C203 and C202. The maximum oscillating current is in the tuned primary circuit of T201, and the resulting voltage from the primary current is taken from the secondary of T201 and used as the input source for IC201A. The input connections are to pins 1 and 2.

The sound IC performs two major circuit operations. The first circuit function is sound-IF amplification. IC201A acts as a sideband IF amplifier and limiter. The B+ voltage is supplied to pin 14 from the +25V source. After amplification in IC201A, the sound signal is fed from pins 11 and 12 to discriminator transformer T202. It is then detected in IC201B and passed via pin 13 through C210 to the volume and tone controls.

Probable Cause: The loss of sound in this case was due to an open winding in transformer T201. For loss of sound, low volume, or distorted sound, the prime suspect would be a defective IC201. Signal injection with a generator, or signal tracing with a high-gain audio amplifier is a quick way to isolate such audio faults. A scope can also be used to trace the signal throughout the various audio stages.

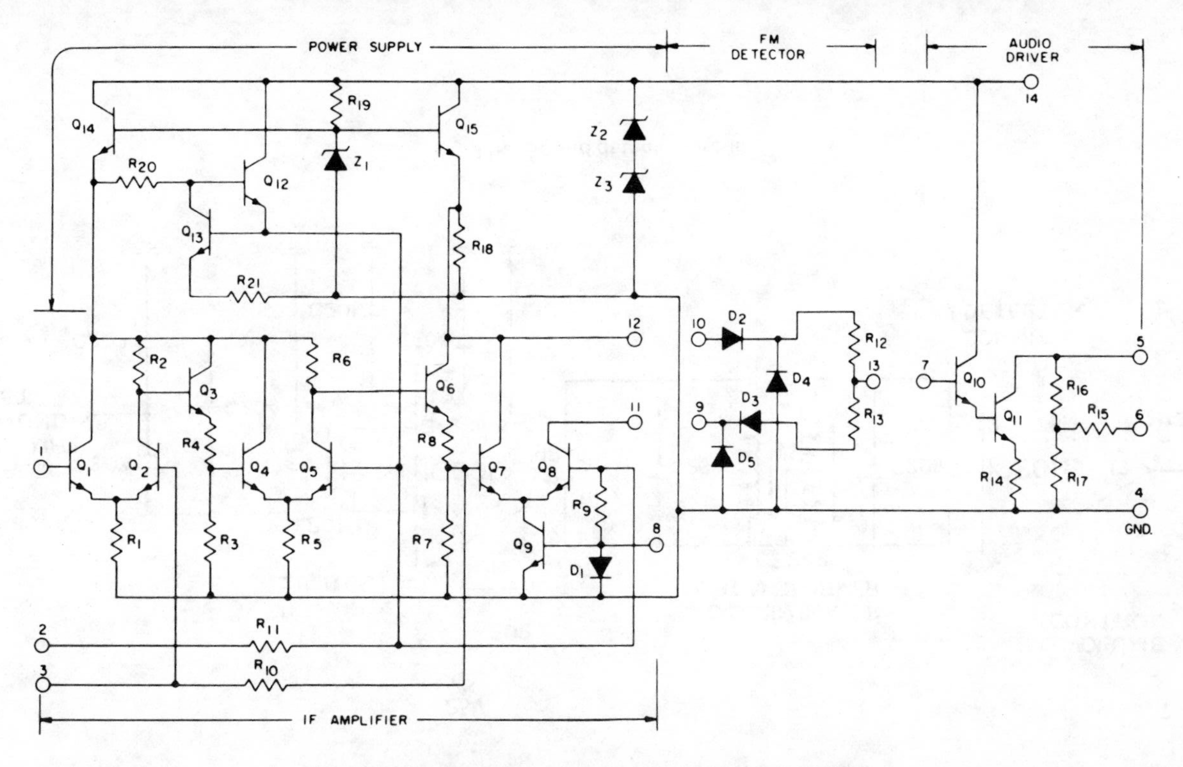

Fig. 11-1. Internal components of integrated circuit IC201.

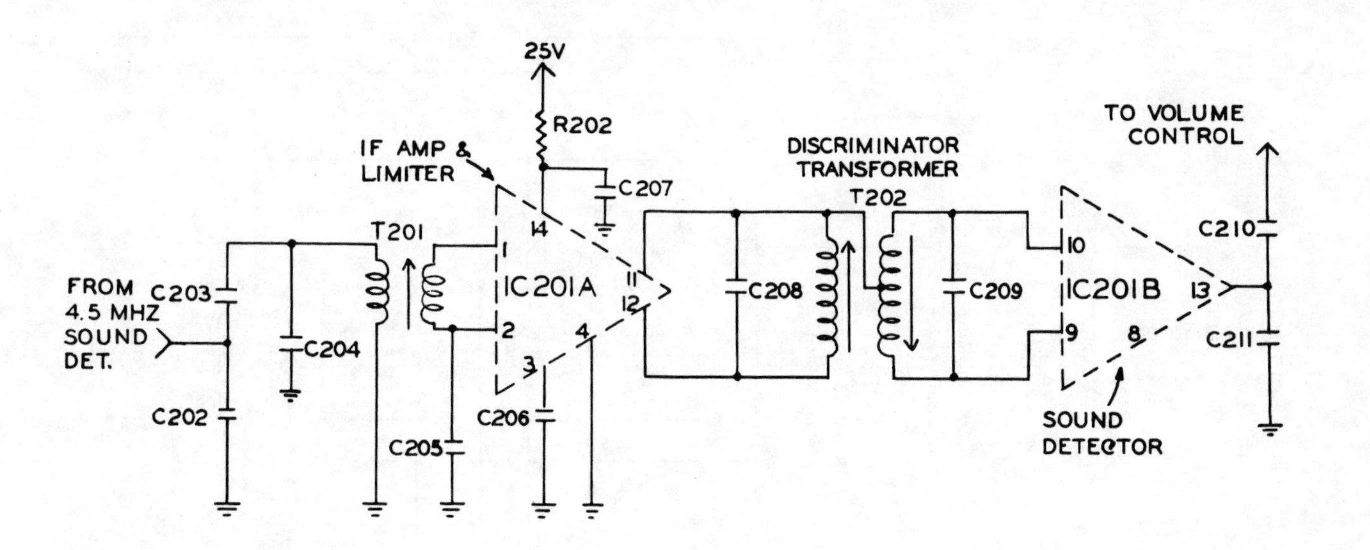

Fig. 11-2. Sound detector circuit.

MOTOROLA TS-931 CHASSIS (QUASAR)

Remote Control: Some remote control functions have erratic operations and respond to spurious noise pulses. Look for trouble in the remote YA panel, noise immunity circuit.

Circuit Operation: The remote control panel that is coded "YA-15" and later has a noise immunity circuit (Fig. 11-3) following each pulse detector. The noise immunity circuit containing transistor Q14 responds to the continuous signal that would be received from the remote transmitter, but discriminates against intermittent noise pulses. The circuit is designed with a slow-charge fast-discharge characteristic. Thus a continuous signal applied to Q14 permits C6 to charge to a sufficient voltage to operate the remote functions, while short bursts of noise are quickly discharged by Q14 before the capacitor can build up a sufficient triggering voltage.

The complete remote control schematic in Fig. 11-4 shows that the noise immunity circuits follow on/off pulse detector Q5 and channel-change pulse detector Q4.

The charging circuit in Fig. 11-3 consists of Q5, D10, and C6 in series across the B+ supply. During the half-cycle of input signal when Q5 is turned on, C6 starts to charge. With a continuous signal present at the anode of D10, Q14 is biased off by the voltage drop across D10, and base-circuit capacitor C20 remains charged, keeping Q14 turned off. Capacitor C6 then charges slowly up to the voltage required to activate the pulse shaper Q6 and Q7.

But when a signal is not being transmitted, C20 can discharge rapidly through R56. This forward-biases Q14,

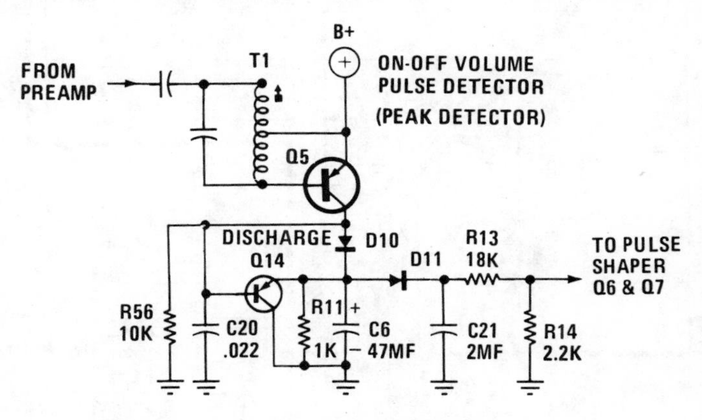

Fig. 11-3. Noise immunity circuit.

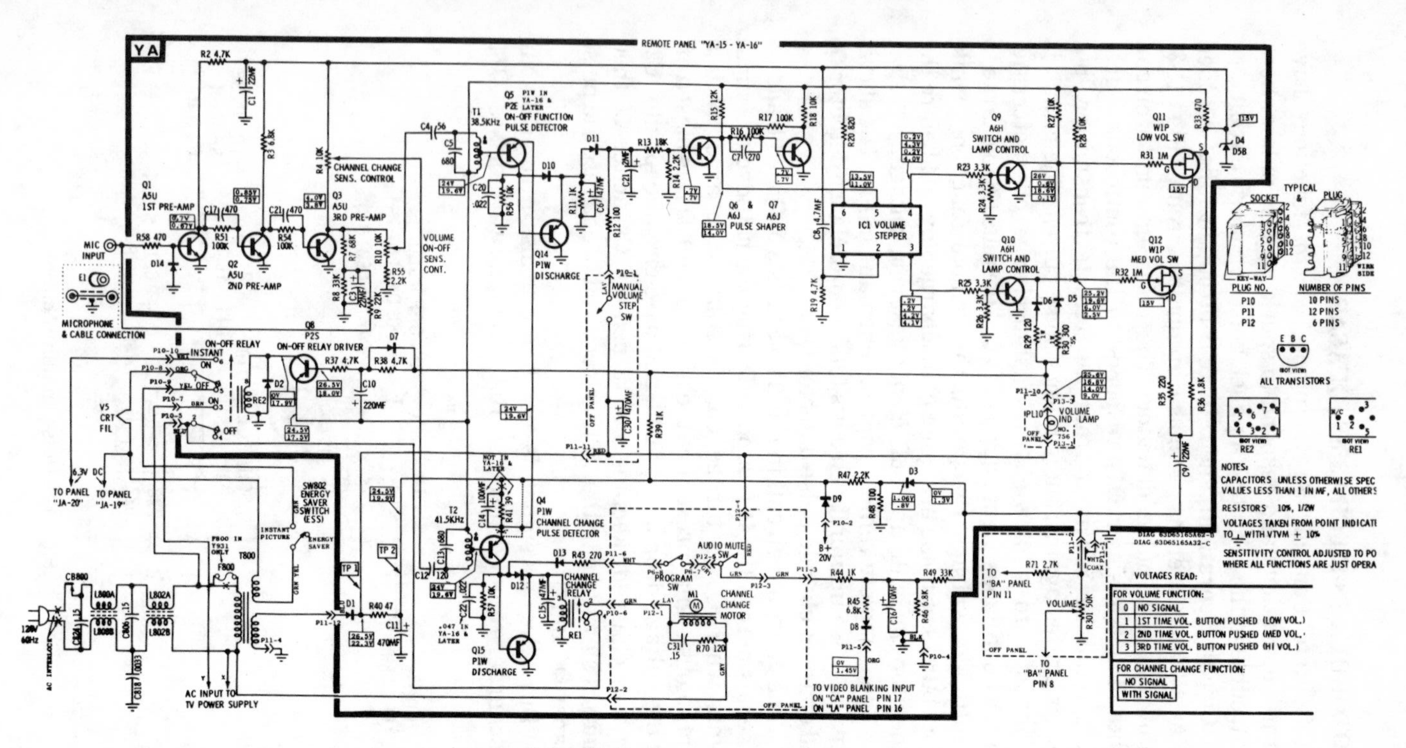

Fig. 11-4. Remote control, YA panel.

making it discharge C6. C6 will then discharge in about 5 milliseconds when the input drops to zero. So when noise is present, this rapid discharge occurs every time there is a "gap" between noise pulses. Therefore C6 does not develop enough charge to activate the shaper pulse circuit.

Probable Cause: The false response of the remote control unit was caused by a shorted diode D10 in the noise immunity circuit. You might also suspect an open C20 capacitor.

ZENITH 17EC45 CHASSIS

Remote Control: Set cannot be turned off with the three-button remote control unit or with the manual off/on button. Look for trouble in the volume stepping and on/off circuits located in the remote control chassis.

Circuit Operation: A 40.25 MHz signal from the hand-held control unit is selected by detector coil transformer L203 in Fig. 11-5, rectified by diode CR206, and coupled to the base of Q201. This voltage causes the Schmitt trigger to change states. Then the Schmitt trigger, a dual toggle multivibrator, and a resistive network, function to provide the proper volume and on/off steps.

With the set off, but with the line cord plugged in, output pins 3 and 4 of IC201 will be low (one volt or less). This is the *TV off* condition. Q203 is not conducting, and the collector voltage is +25V. Snapping the volume button on the hand unit will cause multivibrator IC201 to switch in sequence through the states of voltages at the output pins.

With Q203 cutoff, no current flows through the lamp in A201, the photo-optical isolator. This is an encapsulated unit that includes a lamp and a light-dependent resistor. The resistor has a low value when the light is on, so this device acts as a switch to turn on the gate of triac CR222. It also isolates the control circuit from the triac gate, which has AC line potential on it.

The triac is a solid-state bidirectional switch, conducting current in both directions—similar in operation to two SCRs connected in reverse-parallel. This device has the ability to handle the AC power needs of the TV chassis.

The first snap of the remote unit causes pin 4 to be high (6V or more) and pin 3 to be low (1V or less). The positive voltage at pin 4 causes diode CR217 to conduct and turn on Q203. This activates A201 and fires the triac which turns on the set.

The second and third click of the remote control unit changes the multivibrator state and adjusts the volume. The fourth click of the hand unit returns the multivibrator to the *TV off* condition.

Probable Cause: If the set turns on with the manual switch, but not by remote control, check for a faulty L203 coil or CR206 diode. Coil L203 may only have to be peaked up.

Suspect an open CR218 diode if the set turns off with a volume step. If set will not turn on, check for an open A201 or CR222.

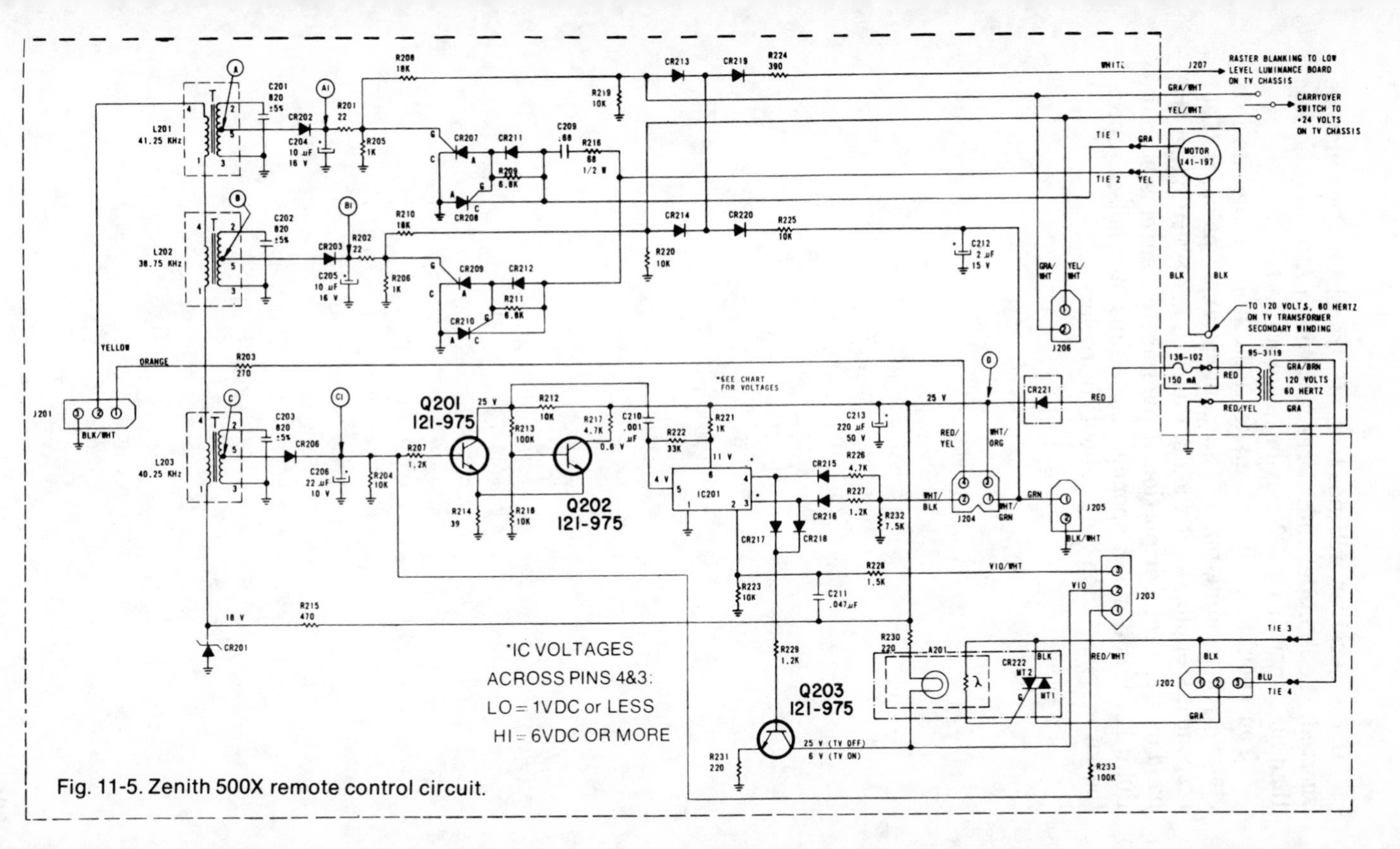

Fig. 11-5. Zenith 500X remote control circuit.

If the receiver will not turn off, transistor Q203 may be shorted, the light-dependent resistor within A201 may be less than 10K, or traic CR222 may be leaky or shorted.

You should be able to make IC201 step through its four states by grounding pin 5 to simulate an input pulse. If IC201 responds, stepping as it should, then you know that either the hand unit or receiver portion of the remote control is at fault. But if IC201 does not respond to these simulated pulses (check the output pins), then the IC should be replaced.

ZENITH 25FC45 AND OTHER F-LINE CHASSIS

Remote Control: TV receiver could not be turned on. Look for trouble in the following remote-control-system functions (note Fig. 11-6):

- Remote power supply
- Triac
- Switching transistor
- Photo-optical isolator
- Multivibrator
- Schmitt trigger
- Microphone and amplifier
- Hand control unit

Circuit Operation: As a preliminary step, check the AC line cord and reset the circuit breaker. Next, operate the manual power on/off switch on the front panel. If the set comes on, then the hand transmitter, detector circuit, or microphone/amplifier is at fault. If the set will not turn on with the manual button, one of the functional blocks between the detector circuit and power supply would be faulty.

To find out which section is defective, the following checks can be made. Refer to the schematic in Fig. 11-7.

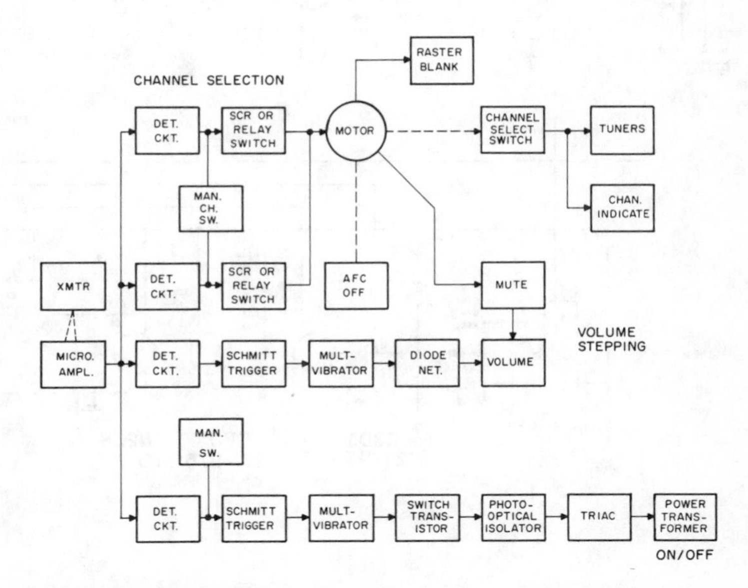

Fig. 11-6. Remote control function, block diagram.

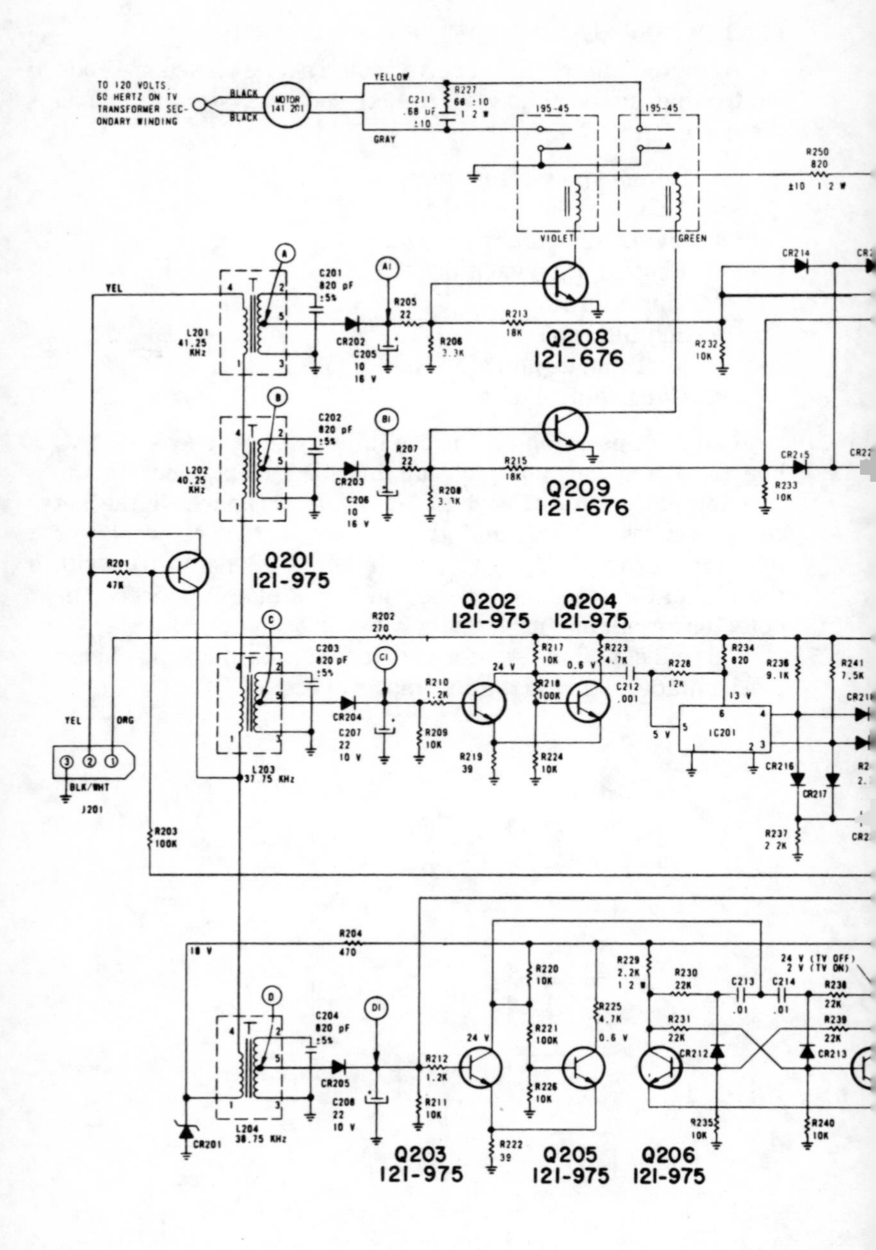

TO 120 VOLTS, 60 HERTZ ON TV TRANSFORMER SECONDARY WINDING
BLACK
BLACK
MOTOR 141 201
YELLOW
GRAY
C211 .68 uF ±10
R227 68 ±10 1 2 W
195-45
195-45
VIOLET
GREEN
R250 820 ±10 1 2 W
CR214
Q208 121-676
Q209 121-676
Q201 121-975
Q202 121-975
Q204 121-975
Q203 121-975
Q205 121-975
Q206 121-975
L201 41.25 KHz
L202 40.25 KHz
L203 37 75 KHz
L204 38.75 KHz
C201 820 pF ±5%
C202 820 pF ±5%
C203 820 pF ±5%
C204 820 pF ±5%
CR202
CR203
CR204
CR205
CR201
C205 10 16 V
C206 10 16 V
C207 22 10 V
C208 22 10 V
R205 22
R206 3.3K
R207 22
R208 3.3K
R213 18K
R215 18K
R232 10K
R233 10K
CR215
R201 47K
R202 270
R203 100K
R204 470
18 V
YEL
ORG
BLK/WHT
J201
R210 1.2K
R209 10K
R217 10K
R218 100K
R219 39
R224 10K
R223 4.7K
24 V
0.6 V
C212 .001
R228 12K
R234 820
13 V
5 V
IC201
R236 9.1K
R241 7.5K
CR216
CR217
R237 2 2K
R212 1.2K
R211 10K
R220 10K
R221 100K
R226 10K
R222 39
R225 4.7K
0.6 V
24 V
R229 2.2K 1 2 W
R230 22K
R231 22K
C213 .01
C214 .01
24 V (TV OFF)
2 V (TV ON)
R238 22K
R239 22K
CR212
CR213
R235 10K
R240 10K

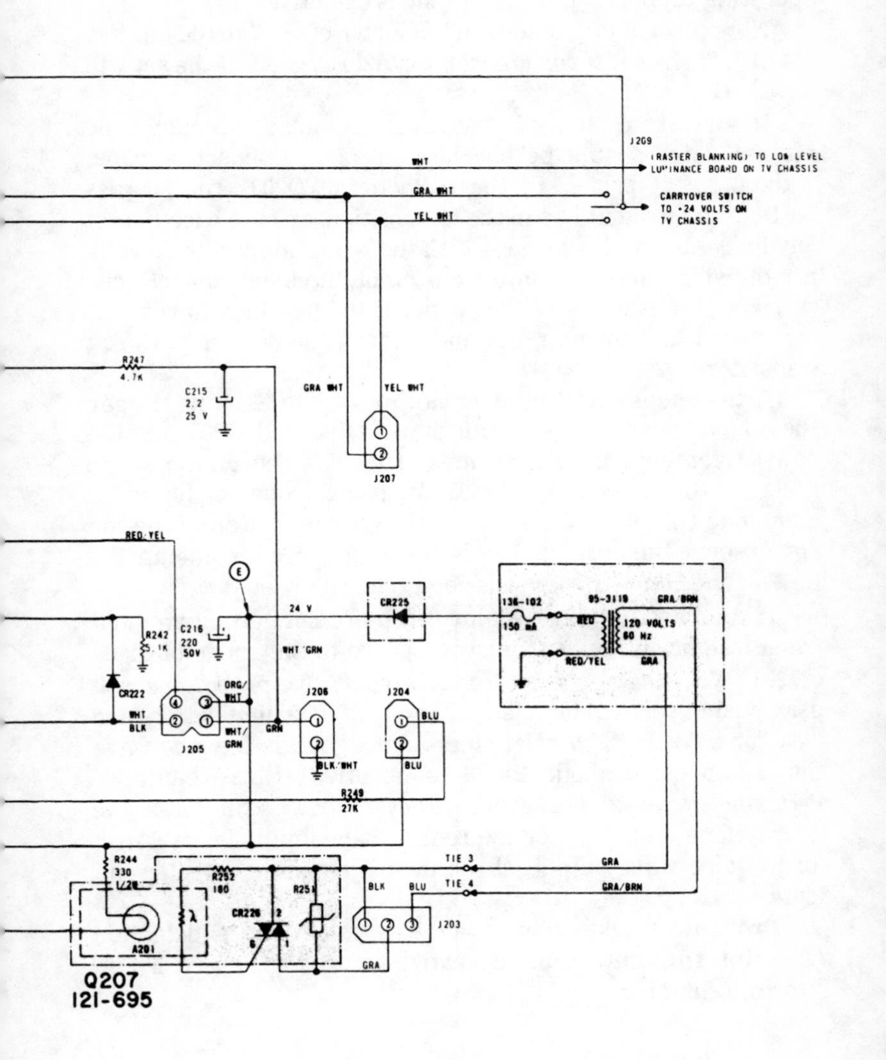

Fig. 11-7. Zenith 600Z remote control chassis.

Place a 680Ω resistor across the light-dependent resistor leads of isolator A201. If the set turns on, triac CR226 is operating properly. If not, the triac is defective.

Now place a clip lead from collector of Q207 to the emitter of Q207. If the set turns on, isolator A201 is good. If the set will not turn on, A201 is faulty.

If the traic and photo-optical isolator test okay, the multivibrator should be checked out next. Connect a scope, with a direct probe, to the collector of Q203. The scope's vertical input should be in the DC position, and the baseline set for the center of the screen. With the scope gain set at 10 volts per division, snap the on/off button on the hand control unit. Observe the baseline of the scope. If the baseline moves, yet the set will not turn on, the multivibrator is defective. Check transistors Q206 and Q207.

If the scope baseline does not move, the Schmitt trigger should be checked out. Transistors Q203 and Q205 are the active elements in the Schmitt trigger. Connect the scope probe to the cathode of CR205. With the scope set for a DC input, one volt per division vertical gain, snap the on/off button and observe the base line. If the baseline moved momentarily, R212 or the Schmitt trigger is defective.

If the baseline did not move, you are not getting an input trigger signal. Connect the scope probe to the anode of diode CR205. With the scope set for AC input, 10X probe, one volt per division vertical gain, snap the on/off hand unit button, and look for a 5V P-P, 40 kHz, sine-wave signal. If the sine-wave signal is present, diode CR205 is defective. If less than a 3V P-P sine wave is observed, either detector coil L204, the microphone amplifier, or the remote hand unit is faulty. Also, for low sine-wave output, check the alignment of detector coil L204.

Probable Cause: This fault was caused by a defective A201. Intermittent on/off operation is usually caused by an intermittent triac.

ZENITH G-LINE CHASSIS

Remote Control: Receiver's remote control functions will not operate. The electronic hand transmitter does not seem to be working.

Circuit Operation: Refer to Fig. 11-8 for the schematic of the SC1000 electronic transmitter. The frequency of the oscillator is varied by changing the ratio of the inductance and capacitance in the emitter circuit of Q2. This ratio change is accomplished by the closing of switches S1 through S6. Pushing switch S1 develops the highest frequency, and S6 will produce the lowest frequency. These six frequencies range from 37.86 to 42.66 kHz. Two adjustments, L1 and R12, set all six frequencies—when the high and low frequencies are set, the other frequencies will be correct.

The oscillator output signal is amplified by Q3, the output stage. This stage drives the ultrasonic speaker that emits a signal directed at the microphone in the receiver. The output signal of the electronic transmitter is larger in amplitude than the output of previous mechanical-rod transmitter units.

The most common problems with the remote transmitter are no signal output, very low output, or an off-frequency condition. Use an oscilloscope to trace through the oscillator and output driver stages. With a scope connected across the ultrasonic speaker, a 20V to 50V P-P signal (Fig. 11-9) should be seen for a properly operating unit. Also check for low battery, poor pushbutton contacts, and faulty transistors.

For frequency alignment, connect the scope leads across the speaker terminals and set the time base to 10 microseconds/division. When a function button is pushed, the scope waveform should have at least a 20V P-P signal. A frequency counter must be used to set the frequencies to within 20 Hz. Connect the frequency counter across terminals D43 and H43 of the 9-127-02 remote module located on the main chassis. If adjustment is necessary , adjust coil L1 to set the lowest command frequency, which activates the mute function at 37.86 kHz. Next, adjust R12 to set the highest frequency, for the zoom function at 42.66 kHz. These two controls will interact and should be adjusted a number of times until both frequencies are correct.

Probable Cause: This symptom was caused by a defective (open) ultrasonic speaker.

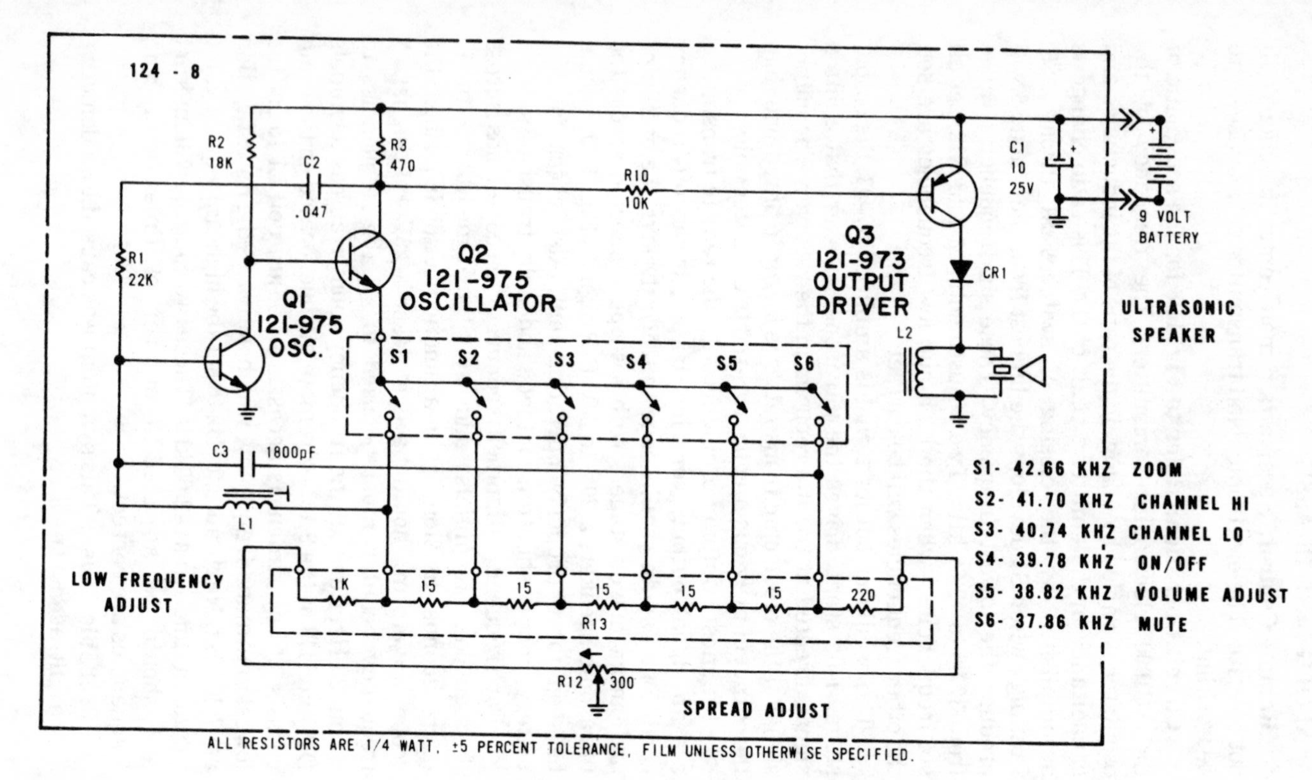

Fig. 11-8. Schematic of SC1000 remote transmitter.

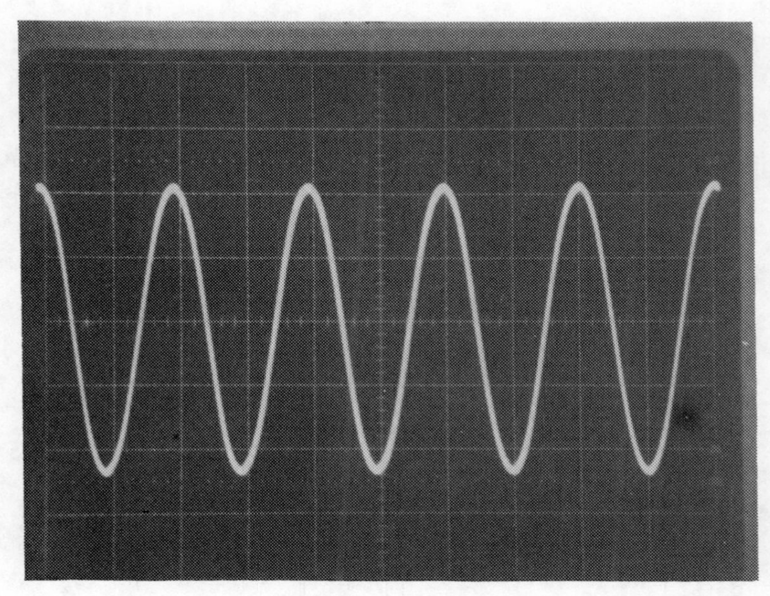

Fig. 11-9. Correct 20V to 50V P-P signal at ultrasonic speaker terminals.

Index

Index